New Directions in the Philosophy of Science

Series Editor: **Steven French**, Philosophy, University of Leeds, UK

The philosophy of science is going through exciting times. New and productive relationships are being sought with the history of science. Illuminating and innovative comparisons are being developed between the philosophy of science and the philosophy of art. The role of mathematics in science is being opened up to renewed scrutiny in the light of original case studies. The philosophies of particular sciences are both drawing on and feeding into new work in metaphysics and the relationships between science, metaphysics and the philosophy of science in general are being re-examined and reconfigured.

The intention behind this new series from Palgrave Macmillan is to offer a new, dedicated, publishing forum for the kind of exciting new work in the philosophy of science that embraces novel directions and fresh perspectives.

To this end, our aim is to publish books that address issues in the philosophy of science in the light of these new developments, including those that attempt to initiate a dialogue between various perspectives, offer constructive and insightful critiques, or bring new areas of science under philosophical scrutiny.

Titles include

THE APPLICABILITY OF MATHEMATICS IN SCIENCE:
Indispensability and Ontology
Sorin Bangu

PHILOSOPHY OF STEM CELL BIOLOGY:
Knowledge in Flesh and Blood
Melinda Fagan

SCIENTIFIC ENQUIRY AND NATURAL KINDS:
From Planets to Mallards
P.D. Magnus

COUNTERFACTUALS AND SCIENTIFIC REALISM
Michael J. Shaffer

MODELS AS MAKE-BELIEVE:
Imagination, Fiction and Scientific Representation
Adam Toon

Forthcoming titles include:

THE PHILOSOPHY OF EPIDEMIOLOGY
Alex Broadbent

SCIENTIFIC MODELS AND REPRESENTATION
Gabriele Contessa

CAUSATION AND ITS BASIS IN FUNDAMENTAL PHYSICS
Douglas Kutach

BETWEEN SCIENCE, METAPHYSICS AND COMMON SENSE
Matteo Morganti

ARE SPECIES REAL?
Matthew Slater

THE NATURE OF CLASSIFICATION
John S. Wilkins and Malte C. Ebach

New Directions of the Philosophy of Science
Series Standing Order ISBN 978-0-230-20210-8 (hardcover)
(*outside North America only*)

You can receive future titles in this series as they are published by placing a standing order. Please contact your bookseller or, in case of difficulty, write to us at the address below with your name and address, the title of the series and the ISBN quoted above.

Customer Services Department, Macmillan Distribution Ltd, Houndmills, Basingstoke, Hampshire RG21 6XS, England

Scientific Enquiry and Natural Kinds

From Planets to Mallards

P. D. Magnus
University at Albany, SUNY, USA

© P. D. Magnus 2012

All rights reserved. No reproduction, copy or transmission of this publication may be made without written permission.

No portion of this publication may be reproduced, copied or transmitted save with written permission or in accordance with the provisions of the Copyright, Designs and Patents Act 1988, or under the terms of any licence permitting limited copying issued by the Copyright Licensing Agency, Saffron House, 6–10 Kirby Street, London EC1N 8TS.

Any person who does any unauthorized act in relation to this publication may be liable to criminal prosecution and civil claims for damages.

The author has asserted his right to be identified as the author of this work in accordance with the Copyright, Designs and Patents Act 1988.

First published 2012 by
PALGRAVE MACMILLAN

Palgrave Macmillan in the UK is an imprint of Macmillan Publishers Limited, registered in England, company number 785998, of Houndmills, Basingstoke, Hampshire RG21 6XS.

Palgrave Macmillan in the US is a division of St Martin's Press LLC, 175 Fifth Avenue, New York, NY 10010.

Palgrave Macmillan is the global academic imprint of the above companies and has companies and representatives throughout the world.

Palgrave® and Macmillan® are registered trademarks in the United States, the United Kingdom, Europe and other countries.

ISBN 978–0–230–36917–7

This book is printed on paper suitable for recycling and made from fully managed and sustained forest sources. Logging, pulping and manufacturing processes are expected to conform to the environmental regulations of the country of origin.

A catalogue record for this book is available from the British Library.

A catalog record for this book is available from the Library of Congress.

10 9 8 7 6 5 4 3 2 1
21 20 19 18 17 16 15 14 13 12

Printed and bound in the United States of America

Contents

List of Illustrations   viii

Acknowledgments   ix

Series Editor's Foreword   xi

Introduction   1

1   **How to Think about Natural Kinds**   4
    A.   Why history is no help   5
    B.   Some criteria considered   7
       B.1   The induction assumption   8
       B.2   The essence assumption   18
       B.3   The science assumption   19
       B.4   The law assumption   21
       B.5   Artifacts and artificial kinds   23
       B.6   The sharpness assumption   26
       B.7   Starting with language   29
       B.8   The intrinsic feature assumption   32
       B.9   The hierarchy assumption   37
       B.10   The scarcity assumption   38
       B.11   The implicit simpliciter assumption   39
    C.   Keeping score   45

2   **A Modest Definition**   47
    A.   First formulation   47
    B.   More or less natural kinds   49
       B.1   Lessons from underdetermination   50
       B.2   The lessons applied   52
    C.   Induction redux   55
    D.   Natural kinds for settled science   57
       D.1   Example: the domain of chemistry   57
       D.2   Fungible kinds   61

3	**Natural Kinds Put to Work**	68
A.	Eight planets, great planets	68
	A.1 Numerology and asteroids	70
	A.2 Enter Pluto	71
	A.3 The constraints of astronomy	73
	A.4 Natural kinds and the fate of Pluto	78
B.	The abundance of living things	83
	B.1 Particular species (buzz, buzz)	85
	B.2 The species category	87
	B.3 How species are and are not natural kinds	94
C.	Thinking outside the box	96
	C.1 Tasks and processes	96
	C.2 Distributed cognition	97
	C.3 What the natural kind is not	101
D.	Further examples	102

4	**Practical and Impractical Ontology**	103
A.	An unreasonable dichotomy	103
	A.1 Natural kinds and bicameral legislation	105
	A.2 Amphibolic pragmatism	109
	A.3 The pragmatists' hope of convergence	113
	A.4 Engaging the world	115
	A.5 The practical as leverage on the real	117
B.	Deep metaphysics	119
	B.1 Bad arguments for realism	120
	B.2 Realism and metaphysical depth	123

5	**The Menace of Triviality**	126
A.	Cheap similarity	127
B.	Project-relative kinds	129
	B.1 Promiscuous realism	130
	B.2 Cooking up natural kinds	133
C.	Agent-relative kinds	136
	C.1 Meerkat threats and alarms	137
	C.2 Unicorns and fictobiology	140
	C.3 Constellations	142
D.	Coda on promiscuity	145

6	**Causal Processes and Property Clusters**	**147**
	A. Species as the specimen of an HPC	149
	A.1 Worries about polymorphism	151
	A.2 Getting over similarity fetishism	156
	A.3 Natural kinds and systematic explanation	160
	B. Species and token histories	165
	B.1 The tigers of Mars	166
	B.2 Hybrids and separate origin	168
	C. RE: ducks, the species problem redux	172
	D. Historical individuals	175
	D.1 Sets and sums, a metaphysical non sequitur	176
	D.2 Metaphysical puzzles about change	181
	D.3 Individualism and HPCs	182
	E. HPC thinking beyond token causes	183
	E.1 The waters of Mars	184
	E.2 The Unity Problem	188
	F. Coda on HPCs	190
7	**Conclusion**	**192**
Bibliography		195
Index		207

List of Illustrations

1.1	Projectible predicates for clearly ordered kinds	13
1.2	Projectible predicates for unordered kinds	15
2.1	Projectible predicates in a restricted domain	56
3.1	Estimates of Pluto's mass	73
3.2	Our solar system	74
3.3	Lagrange Points	76
3.4	A cladogram of anglerfish	90
4.1	Boyd's idiom	107
4.2	Three views about natural kinds	117
6.1	Female *Linophryne lucifer*	159
6.2	Female and male *Linophryne arborifera*	161
6.3	Sets versus fusions	179

Acknowledgments

This book took shape over the course of a semester, while I was a visiting fellow at the Center for Philosophy of Science in Pittsburgh. Yet the ideas are ones that I have been thinking about for a long time, and so I have a long list of intellectual debts.

I began thinking about natural kinds in the last century, when I was a graduate student in San Diego. I studied with Philip Kitcher, when he was reconsidering the view he had defended in *The Advancement of Science* (1993) and working toward *Science, Truth, and Democracy* (2001). I urged him to be more of a pragmatist, and he urged me to be more of a realist. It is possible that we have switched places. When Philip left UCSD, I was advised first by Sandra Mitchell and later by Paul Churchland. My dissertation was on the underdetermination of theory by data, and worrying about underdetermination kept me busy for several years.

I still thought about realism and natural kinds off and on. At the Philosophy of Science Association meeting in 2004, I had a late night conversation about natural kinds in the hotel bar with Richard Boyd, John Dupré, Jay Odenbaugh, Michael Weisberg, and Robin Hendry. Not too long after that, I started a blog. I made a series of posts in 2006 about the relation between realism and ontological pluralism. I had a rewarding conversation in the comments section with Jay Odenbaugh, Matt Brown, and Greg Frost-Arnold.

In Spring 2010, I was invited to give a talk at Cornell University. They encouraged me to present work-in-progress, so I gave a paper trying to develop my still half-baked thoughts on natural kinds. I thank the helpful and indulgent audience, especially Adam Bendorf, Karen Bennett, Matti Eklund, and Nico Silins.

In Fall 2010, I had just gotten tenure at the University at Albany, and I spent the semester at the Center for Philosophy of Science. There I began to make a thorough, systematic slog through the literature. I met lots of people in Pittsburgh, and they would ask politely what I was working on. I ruthlessly exploited this opportunity to try out a succession of possible positions. Showing up to my Center office every morning, I began writing on natural kinds in earnest.

Notes grew to the size of papers, papers grew to the size of chapters, and without planning to be I was on the way to writing a book.

I benefited from interacting with all of the philosophers who were visiting the Center that semester. I extend special thanks to Heather Douglas, Kareem Khalifa, Bert Leuridan, Richard Samuels, and Peter Vickers. Thanks are also due to many members of the Pittsburgh philosophical community who gave me feedback on parts of the project, especially Jim Bogen, Julia Bursten, David Danks, Eduoard Machery, Sandy Mitchell, and Jim Woodward. A special shout out is due John Norton, then director of the Center, who was a key organizer for lots of interactions and was himself a smart interlocutor. I gave a talk as part of the Center's lunchtime colloquium series which covered what became the first chapters of the book.

The discussion of planets (Chapter 3 § A) was presented to the visiting fellows reading group in Pittsburgh and delivered as a talk at City College of New York in November 2010.

The discussion of ducks and seadevils (Chapter 6 § A) was presented to the reading group in Pittsburgh, presented at Metaphysics & Philosophy of Science in Toronto, May 2011, and published in *Philosophy and Biology* (Magnus 2011a). Thanks to Jonathan Birch and Joseph McCaffrey for pointing me toward interesting biological examples.

Thanks to Cristyn Magnus, who discussed many of the ideas as I mulled them over, read fragments of drafts, and encouraged me to actually pull it together. Thanks to my colleagues who I harangued about natural kinds over the years, especially Ron McClamrock and Brad Armour-Garb. Ingo Brigandt read a draft of the book and provided many helpful comments.

I used the manuscript as part of a graduate seminar at the University at Albany in Fall 2011. Thanks to the participants for their contributions, especially to Brian Deinhart, Chris DeLeo, Daniel Feuer, John Milanese, Tiffany Redies, and Damian Thibeault.

This attempt to trace out influences on my thinking no doubt leaves out profound ones, and I will inevitably fail to acknowledge people whom I surely ought to thank. Thanks to all of them, too.

The material in Chapter 6 § A was previously published in *Biology & Philosophy* © Springer Science+Business Media.

Most of the figures are my own illustrations. Exceptions are indicated in the figure captions. Figure 3.1 is used with permission from the American Geophysical Union. Figures 3.4 and 6.2 are free to use under a Creative Commons license.

Series Editor's Foreword

The intention behind this series is to offer a new, dedicated, publishing forum for the kind of exciting new work in the philosophy of science that embraces novel directions and fresh perspectives. To this end, our aim is to publish books that address issues in the philosophy of science in the light of these new developments, including those that attempt to initiate a dialogue between various perspectives, offer constructive and insightful critiques, or bring new areas of science under philosophical scrutiny.

P. D. Magnus has written just the kind of vibrant and provocative work that we were hoping to attract when the idea of the series was first mooted. It offers a crisp, fresh approach to a well-known issue in the philosophy of science – namely our understanding of natural kinds and the form of realism that best accompanies that understanding. Here Magnus sees the crucial issue as having to do with scientific taxonomy and how the categories that are posited fit the world. Among the many virtues of the book are some beautifully presented and illuminating case studies, from, as the title suggests, planets (with a timely discussion of the fate of Pluto) to mallards, peppered moths and anglerfish. By starting with this perspective he hopes to avoid becoming entangled with the semantic issues or deep metaphysics that have bedevilled previous accounts. In his claim that a category of things or phenomena is a natural kind for a domain if it is indispensable for successful science in that domain, Magnus moves the relevant discussion on from the increasingly moribund concerns with essentialism in this context. And by situating this account in the context of his 'pragmatic naturalism' and 'modest realism', he offers the possibility of revitalising the realism–antirealism debate by bringing the terms of that debate into closer engagement with scientific practice. In this respect, as Magnus himself notes, this work can be aligned with the strong pragmatic traditions that run throughout the philosophy of science. And of course, he does not shy away from critically engaging with the recent literature on natural kinds and in particular with well-entrenched accounts such as that of kinds as 'homeostatic property clusters' which has become the consensus view. As Magnus argues, although this account certainly gains traction in the case of biological species, it faces acute difficulties when it comes to other kinds, not just in physics, but in chemistry as well. Here we get perhaps the most nuanced discussion of these issues currently available. But more significantly, what we have here is a book that takes the debate forward, offers a number of thought-provoking claims and, indeed, presents a new direction in the philosophy of science.

Steven French
Professor of Philosophy of Science
University of Leeds

Introduction

In Plato's *Phaedrus*, Socrates is concerned with how things are grouped together and how they are divided. He makes an analogy with carving up an animal. Just as cuts of meat should be carved at the joints rather than broken across bones, our account of the world should carve nature at its joints. So we inherit this grisly metaphor for what scientific enquiry does when it aims to discover the real divisions in nature: When science finds the natural kinds, its concepts are the chops and steaks of the world.

The metaphor is both grisly and unfortunate. A butcher is only interested in the animal for one project: butchering. Scientific enquiry is not so unified. The pig which is butchered just one way may be described by scientists as consisting of molecules, cells, or metabolic systems, and none of these descriptions is more natural than the others. The problem is magnified when we enlarge our view from the pig and think about the *world* altogether. The problem is not that there are no rewarding places to cut, but that there are too many. There are so many joints in the world that we could not possibly carve it up along all of them. This might be a cause for nominalism or despair, but it need not be. The divisions we find might be *real* features of the world, even if they are more numerous than those split by the butcher. It is an important fact about the pig's biology that it is made up of cells, and an important fact of its biochemistry that it is made up of molecules. No single enquiry can make all the right cuts. Rather, different enquiries require cutting along different joints.

Let's leave the ancient butchershop and consider the night sky. Even Socrates knew about planets like Mars and Jupiter. There are others which he did not know about: Uranus, Neptune. And there is a dim point in the sky – Pluto – which I was taught about as a planet but which astronomers no longer consider to be one. In 2006, when astronomers regimented the use of the world 'planet' in a way that excluded Pluto, were they carving the solar system at a joint? Is the **planet** category a natural kind?

Questions like this can be raised about any of the categories that are employed in scientific accounts of the world. As a philosopher of science, I want the resources to think about such questions.

Of course, *natural kind* talk brings with it a train of philosophical baggage. In recent decades, the issue has mostly been treated as a matter of natural kind *terms*, the language that we use to pick out categories. This makes it an issue about nomenclature and about the way that language works. Applying such an approach to the case of Pluto, the issue of whether there is a natural kind structure to the solar system becomes entangled with the semantics for the word 'planet' as it was used by astronomers at various times. I suggest we start somewhere else – asking instead about taxonomy, the categories we posit, and how they fit the world. This alternative starting place allows us to identify natural kinds without pretending to have a complete account of semantics or deep insight into fundamental metaphysics. The resulting account of natural kinds is modest and pragmatic, but nonetheless realist.

To put it loosely, the account of natural kinds which I defend maintains that a category of things or phenomena is a *natural kind* for a domain if it is indispensable for successful science of that domain. Scientific success involves making sense of the things or phenomena – both accurately predicting what they will do and explaining their features. The account conflicts with the tradition that associates natural kinds with fundamental and precise *essences*.

I claim no great originality in associating natural kinds with conditions for inductive and explanatory success. The connection to successful science echoes many voices in the pragmatist tradition. Although pragmatism is sometimes taken to be anti-realist or even idealist – even by some pragmatists themselves – I argue that the natural kinds that I describe are *features of the world* in a straightforward way. The appropriate metaphysical picture is one of pragmatic naturalism and modest realism.

Ultimately, an account of natural kinds must pay its rent by helping us understand actual science and its relation to the world. So I discuss both a large number and wide range of examples. Some of these, like chemical kinds and species, are well-explored terrain. Some of the specific cases, like gold, water, and tigers, are much interrogated denizens of that terrain. Yet I try to avoid the uninformed casual gesture to tigers or lemons. My arguments will range over mosquitoes, ducks, and deepsea fish; the details turn out to matter. I also consider cases, like the planets, which have received little philosophical attention.

In Chapter 1, I make a survey of things that people have thought about natural kinds. I argue for the centrality of three constraints: that natural kinds should support induction, that natural kinds should figure in successful science, and that natural kinds are relative to domains of enquiry. In Chapter 2, I offer my account as a promising way of satisfying these three constraints. In Chapter 3, I apply the account to several specific examples: planets, species, and distributed cognition. In Chapter 4, I articulate pragmatic naturalism and clarify the sense in which my account is realist. In Chapter 5, I entertain the objection that my account is too liberal and includes too many categories as natural kinds. In answering the objection, I consider examples of culinary taxonomy and animal signaling. In Chapter 6, I consider the relationship between natural kinds and homeostatic property clusters.

1
How to Think about Natural Kinds

In many scientific domains, the world constrains our taxonomy. We cannot approach phenomena using just any categories and expect to achieve predictive and practical success. Where our choice is strongly constrained, it is tempting to say that our categories correspond to structures in the world. I think it makes sense to use the label 'natural kind' to pick out such structures.

Other philosophers may reserve the label for something more rarefied, but this threatens to just be a matter of terminology. Something will have gone wrong if it becomes just a fight over who gets the flag of natural kind to stick in the soil of their preferred philosophical conception. One can avoid this outcome by reading everything I have to say as being about *schnatural schkinds*, a category that I am concerned with that I discuss using the words 'natural kinds'. One might even imagine subscripts throughout the text, so that mostly I am on about natural-kinds$_{Magnus}$, which are distinct from natural-kinds$_{Mill}$ and natural-kinds$_{Goodman}$. Although a common practice among philosophers, this kind of insulation by subscripts is dangerous. If we unreflectively add subscripts for every author's characterization of everything, regress threatens. When I write about Hacking's discussion of Mill's account of kinds, should I write natural-kinds$_{Mill_{Hacking}}$? And if other philosophers were to write about my having done so, should they then write natural-kinds$_{Mill_{Hacking_{Magnus}}}$?

In the end, I hope to distinguish something in the neighborhood of natural kinds that is worth thinking about. And I will call these natural kinds.

As an aside: I use boldface type to distinguish **kinds** and **natural kinds** from mere aggregates of things. So dogs are all individual organisms, but **dog** is a kind that includes those organisms as members. The difference should also be clear from context, and the typographic convention is simply meant to underscore the distinction.

A. Why history is no help

We should acknowledge at the outset that 'natural kind' is a term of philosophical jargon. We cannot start from a pretheoretical concept of *natural kind* and provide an analysis of it. A modern Socrates would learn nothing by asking some unsuspecting fellow in the agora what a natural kind is. The words 'natural' and 'kind' can be concatenated to have a sensible English meaning, of course, but not one that philosophers typically intend. We might instead ask what philosophers mean when they use 'natural kinds' as a fixed phrase, but there is no univocal answer to this either. The idea of natural kinds has several sources. Let's look at some of these strands to highlight the difficulty.

The phrase was used by some philosophers before it became jargon. For example, Thomas Reid discussed signs which suggest the signified thing automatically, that 'conjure it up, as it were, by a natural kind of magic' (Reid 1997 [1764], ch. 5, § 3). He meant nothing technical by this, and the phrase is used in a similar casual way by other authors. Some philosophers used the phrase 'natural kind' to describe Platonic forms (e.g. Chretien 1848, p. 77; Martineau 1859, p. 483). None of this was systematic. *Natural kind* only becomes a technical notion after John Stuart Mill's *Logic*.

Mill, following William Whewell, spoke of *Kinds*. John Venn (1866) discusses Mill's view using the words 'natural kinds'; Hacking (1991*b*) thus credits Venn with coining the phrase. There was some subsequent discussion of Mill's view in articles with titles like 'On the doctrine of natural kinds' (Towry 1887) and 'Mill's natural kinds' (Franklin and Franklin 1888), but Mill's account was not widely accepted and the phrase fell out of use. There is no entry for 'natural kind' in James Mark Baldwin's *Dictionary of Philosophy and Psychology* (1901). The entry on 'kind', written by C. S. Peirce, discusses Mill's notion of real kinds without the phrase 'natural kinds'. *Natural kinds* are discussed by Russell (1948, pt. 6, ch. 3) and Quine (1969*a*) in something like Mill's sense, but only to dismiss them. So there is a case to be made that there is no direct trajectory from Mill's Kinds to the more recent vogue of natural kinds. (For more on the substance of Mill's view, see § B.1 below.)

This rise in talk about 'natural kinds' is largely explained by their connection to the Kripke–Putnam account of reference. Peirce, Russell, and Quine had all insisted that natural kinds are prescientific and that part of the business of science is to eliminate them. When Hilary Putnam (1975a [1970]) first appeals to natural kinds to solve problems in semantic theory, however, he specifically invokes them as whatever it is that science tells us are the appropriate categories. A natural kind will have an *essential nature* which we only learn about through scientific enquiry – whatever microstructures are rigidly designated by natural kind terms. Putnam's work was in roughly the same period as Quine's, but Putnam does not wrestle with the incongruity between natural kinds as he invoked them and the nineteenth-century tradition which Quine rejects. (For more about natural kinds and Kripke–Putnam semantics, see § B.7 below.)

Somewhere along the way, Plato's metaphor of carving nature became emblematic of natural kinds. Although there were philosophers before Mill who were happy to use the phrase to describe Platonic forms, Mill and Putnam invoked natural kinds in response to specific philosophical problems. Hacking (1991a) argues that Mill's conception was a response to problems which arose first in the nineteenth century so that it should not be seen as answering ancient questions (see also McOuat 2009). Regardless, the streams have now crossed. There are present philosophers who appeal to ancient views as the way forward in thinking about natural kinds. For example, Brian Ellis (2001) defends a philosophy of science which is essentialist and Aristotelian. Even if Hacking and McOuat are right about the historical contingency of the Millian tradition, Ellis' view is recognizably about *natural kinds* in a contemporary sense.

There is a Lockean tradition which worries about essences in a different way. For Locke, the question is whether the nominal essences of ideas correspond to real essences of things. It is a vexed question how much Locke's position is compatible with more recent views about natural kinds (Mackie 1974; Ayers 1981). Regardless, Locke has been inspirational for contemporary thinkers like Richard Boyd (1991) and Hilary Kornblith (1993). Although they do not make the connection, they could perhaps trace these themes from Locke through Mill and his nineteenth-century critics.

Part of the explosion in 'natural kind' talk since 1970 is unrelated to developments in philosophy. The phrase also appears in psychological literature. Psychologists studying concepts and classification distinguish natural kinds from artifact categories. (See § B.5 below.)

I could go on, but this brief excursion in philology suffices to show that we cannot extract a coherent, univocal definition of 'natural kind' from history. In the face of this cacophony, Hacking (2007*b*) recommends abandoning the phrase 'natural kind' on the grounds that it is too contentious to be useful. One might make the same complaint about any item of philosophical jargon, however. We face a general problem of philosophical methodology: In making a proposal, is it better to express it in familiar vocabulary or to coin neologisms? Hearing a proposal expressed in familiar vocabulary, an audience may assimilate it to things they already understand – or think they understand – and so either overlook the novelty of the proposal or place it in a context that makes it incoherent. Conversely, hearing a proposal built up out of neologisms can be awkward and can make it sound more alien that it is.

In at least some sense, the labels do not matter. Consider a different example: Whether 'whales are fish' is true depends on what we mean by 'fish'. If Linnaeus had coined a new word, as he did for mammals, then we might still say that 'whales are fish'. In both the actual world and this imagined alternative, however, we need to be able to talk about both the category of **marine life** (including both tuna and whales) and the category of **cold-blooded ichthynes** (including tuna but not whales). How the word 'fish' gets sorted out is not the important thing. We should be more concerned with taxonomy than with nomenclature – more with the categories in our scheme of classification than with which labels which are pinned to the categories. That is the standpoint I mean to adopt here with respect to natural kinds.

B. Some criteria considered

Although there is no consensus about what would make a natural kind *natural*, there are a number of traditional and common assumptions. I begin by offering some of these as a list of adages. If these seem telegraphic, don't worry – I will return and discuss each in more detail.

§ B.1 Natural kinds support inductive inferences.
§ B.2 Natural kinds have essences.
§ B.3 Natural kinds are the appropriate categories for scientific enquiry.
§ B.4 Natural kinds are the relata in laws of nature.
§ B.5 Natural kinds are distinct from artificial kinds.
§ B.6 Natural kinds must have sharp boundaries.
§ B.7 Natural kinds should be understood by way of natural kind terms. Natural kinds allow for direct reference.
§ B.8 Membership in a natural kind must be an intrinsic property. Natural kinds are constitutive rather than functional kinds.
§ B.9 Natural kinds are structured hierarchically.
§ B.10 There is a small, manageable number of natural kinds.
§ B.11 A kind is either a natural kind or it is not, without any further specification required.
A kind is natural *for an enquiry*, rather than natural *simpliciter*.

As a cautionary aside: I am not presuming that **natural kind** is a property cluster characterized by these adages. They are not altogether consistent, and I will not ultimately offer an account that vindicates all of them. My only intention here is to explore the terrain in the area of natural kinds. These adages thus serve as starting points, like promotional advertisements in a guidebook to that terrain.

B.1 The induction assumption

A central assumption about natural kinds – the canonical assumption – is that you can make inductive inferences about them. Let's call this the *induction assumption*. This is shared so widely that any reasonable account of natural kinds must vindicate it. Nevertheless, there are problems with the assumption that make it unfit to serve as a definition. The account I ultimately defend will make sense of how the assumption is true, but it will also avoid the difficulties.

Through the twentieth-century analytic lens, the induction assumption is the claim that natural kind *terms* are ones that can enter into successful inductive inference. To use the jargon introduced by

Nelson Goodman, natural kind terms express projectible predicates. The stock example is 'green', in opposition to the nonprojectible 'grue'. Formulations may differ slightly, but the crucial trick is that 'grue' is disjunctive. Pick some future event – like New Year's 2099 in Central Park – and define '*x* is grue' to mean that *x* is green (if observed before that event) but that *x* is blue (if observed during or after that event). In all observations we have made so far, any grue objects have been green. For some observations we have not made yet, grue objects will be blue. All of the emeralds that we have observed have been green, but this along with the definition of 'grue' means that all the emeralds have also been grue. We infer from observation that all emeralds are green, but not that all emeralds are grue. Formally, there is no difference between the two inferences. So what separates them? Goodman's *new riddle of induction* is just the problem of how we can make one inference but not the other.

Once you learn the trick, it is easy to construct grue-like predicates. For example, the counterpart '*x* is bleen' means that *x* is blue (if observed before the specified point) but that *x* is green (if observed then or later). One may object that grue is not a real property, because 'grue' is defined in terms of green and blue. As Goodman points out, however, this is only the result of our starting vocabulary. If we began with a vocabulary that included 'grue' and 'bleen' rather than with a vocabulary that includes 'green' and 'blue', then we would see things the other way round. If we started that way and introduced the term 'green', it would be natural to define '*x* is green' to mean that *x* appears grue (if the observation is made before the specified point) but otherwise that *x* appears bleen. The only reason that grue looks monstrously unnatural to us is that we think green and blue are the real properties.

As Goodman sees it,

> Inductive rightness requires evidence statements and the hypothesis to be in terms of 'genuine' or 'natural' kinds – [i.e.] to be in terms of projectible predicates like 'green' and 'blue' rather than in terms of nonprojectible predicates like 'grue' and 'bleen.' ... Any feasible justification of induction [requires] distinguishing projectible predicates or inductively right categories from others. (1978, pp. 126–7)

So natural kind terms are the projectible predicates required for successful inductive inference. Natural kinds are just the categories picked out by natural kind terms.

Fixation on the examples of *green* and *grue* is somewhat odd, because nobody thinks that **green things** form a natural kind. Quite the contrary, many authors insist that things of a given color do *not* form a natural kind: Mill contrasts legitimate (natural) kinds like **animals** or **sulphur** with mere classes like **white things** or **red things** (1874, bk. I, ch. vii, § 4, p. 79), and Millikan contrasts the bogus kind **round red object** with legitimate kinds like **gold** and **domestic cat** (1984, p. 278). If one infers that all emeralds are green, then it is **emerald** rather than **green** that is supposed to be the natural kind. Of course, this is not to say that new-riddle worries cannot be raised about emeralds. Goodman himself offers the gerrymandered term 'emerose' which applies to emeralds observed before a certain point and to roses observed later (1983 [1954], p. 74, fn. 10). Yet there is something suspicious about the fact that 'grue' and 'emerose' are interchangeably puzzling for Goodman. A green thing becoming blue is no great oddity. Depending on the item in question it might indicate ripening, a new coat of paint, or just a change in temperature. For an emerald to become a rose, however, is an impossibility. This difference between color categories and object categories is perhaps just the distinction between arbitrary collections and natural kinds. If so, we cannot chart that distinction in terms of projectibility.

Richard Boyd, who retains language of 'projectibility' in relation to natural kinds, talks about projectible *theories* rather than projectible predicates (e.g. Boyd 2010, p. 213). A theory which describes objects as emeroses will be systematically unsuccessful. Of course, this is not something that we can determine just on the basis of syntax. Just as 'green' will look gerrymandered to someone who is used to thinking about grue and bleen, 'emerald' will look gerrymandered to someone who is used to thinking about emeroses and roralds (where 'rorald', the counterpart to 'emerose', applies to roses and then to emeralds). Goodman was correct to insist that inductive rightness is not a purely formal or syntactic matter. As Boyd would insist, scientific inference is a complex arrangement of methods that has been developed over time and is subject to further revision. This systematic focus is absent from discussions of *grue*, so I set it aside for now (see § B.11 and Chapter 4 § A.1).

A significant limitation of fixating on *grue* is that we wind up construing *induction* almost exclusively as a projective inference: inferring from a sample A which is F to the conclusion that As are typically Fs. Goodman shows that this schema is not *formally* correct. It will only hold for carefully chosen As and Fs; i.e. for the projectible ones. Coming at natural kinds in this way leads us to suppose that members of a natural kind are connected by similarity. The reason that this A can be used as a proxy for other As is that they all resemble one another in many respects: their Fness, Gness, and so on. To make it a joke, they must be alike in Gness to be alike in species.

Quine, continuing in this vein, writes, 'The notion of a kind and the notion of similarity or resemblance seem to be variants or adaptations of a single notion. Similarity is immediately definable in terms of kind; for, things are similar when they are two of a kind' (1969a, p. 117). Although he goes on to say that this simple definition is not quite enough, he retains a tight connection between kinds and similarity. As he sees it, enquiry begins with kinds that are organized with respect to surface similarity, according to prescientific, common sense metrics of what counts as similar. Science does not typically traffic in kinds like these. The development of science involves first refining what counts as *similar* and, Quine thinks, ultimately shedding the notion of similarity itself. He concludes:

> In general we can take it as a very special mark of the maturity of a branch of science that it no longer needs an irreducible notion of similarity and kind. It is that final stage where the animal vestige is wholly absorbed into the theory. In this career of the similarity notion ... passing ... from the intuitive phase into theoretical similarity, and finally disappearing altogether, we have a paradigm of the evolution of unreason into science. (1969a, p. 138)

It is true that the development of science might *explain* why some things are similar in superficial respects. For example, animals in a particular species share phenotypic characteristics because of shared developmental mechanisms and history. When we recognize this, the surface features cease to be definitive of the species. I elaborate this point in Chapter 6, but the details are not essential here. It suffices to note that the surface features do not 'disappear altogether'. They are used to identify members of the species, and they may also be the reason that we care to distinguish the species at all. Quine might say that biology is not a *mature science* in the relevant sense, but similar points may be made about the electron. It has a cluster of definitive features: mass, charge, and so on. Electrons have a systematic place in the standard model, as one of the three kinds of lepton. The electron mass, however, is a free parameter in the model. The *theoretical similarity* of electrons does not disappear altogether.

Quine is part of a tradition, going back to Mill, which assumes that membership in the same kind is a matter of having a large number of properties in common. Call this *similarity fetishism*. The yoke of similarity fetishism makes the induction assumption unable to accommodate kinds which are not joined by similarity and thus makes it insufficient to serve as a definition of 'natural kind'.

Note: The *induction* in what I have called the induction assumption is projective induction which generalizes from a sample to the class of similar things. The term 'induction' is used more broadly in other contexts, to mean not just projective induction but any non-deductive form of confirmation; it is, in the broader sense, a synonym for 'ampliative inference'. That broader sense is not what I have in mind when discussing the induction assumption, but is closer to the science assumption which I discuss further on (§ B.3).

Limits and degrees

There is a further problem with the induction assumption. To take a standard example, consider a piece of jade. We determine that it is a silicate of calcium and magnesium – nephrite – but we would be wrong to draw conclusions about the composition of jade *simpliciter*. The word 'jade' picks out two different minerals: jadeite and nephrite. So it is typical to say that **jade** is not a natural kind. The problem is that there *are* general facts about jade. Both varieties are fairly hard minerals, which makes them *inedible* and *suitable for making stone tools*. These and many other predicates are projectible for jade *simpliciter*.

To consider a different case, we can conduct epidemiological studies and extrapolate facts about the virulence of syphilis in human populations. So **human** and **syphilis** are natural kinds. Yet, there are limits to the generalizations we can draw from any sample. Some of this is a matter of precision; we can only collect so much data, and there are many possibilities for error. Epidemiology itself necessarily involves certain difficulties; there may be hidden variables and confounding factors which could only be teased out with controlled experiments. And the generalizations have an implicit limit in them; any claims about the spread of syphilis are meant to apply only to humans under unspecified default conditions – not, for example, to humans dosed up with antibiotics.

One might attempt to reply that these limitations are a matter of degree. So, one might say, projectibility of a predicate and naturalness of a kind should also be a matter of degree.

We can illustrate this idea with some schematic examples. If the predicates F and G are weakly projectible for a kind X but strongly projectible for a different kind Y – assuming that no other predicates are projectible for either – then Y would be *more natural* than X. If F, G, and H are all strongly projectible for a third kind Z, then Z would be more natural than both X and Y. These examples are summarized in Figure 1.1.

In this way, we might try to construe kinds as forming a continuum of naturalness. By picking some threshold, we could even say that any kind at least as natural as *that* should count as a natural kind. For example, let Y be the threshold; X is then not natural enough to be *natural*, while Y and Z are. If such a procedure were possible in general, then the induction assumption plus the choice of some threshold would suffice to define 'natural kind' *simpliciter*.

Paul Griffiths suggests appealing to degrees of naturalness in this way, arguing that we can 'distinguish between kinds of greater or lesser naturalness and hence of greater or lesser theoretical value' (1999, p. 217). As he acknowledges, however, it will not be as clean cut as our schematic examples suggest. Degrees of naturalness do not

		Predicates	
		weakly projectible	strongly projectible
Kinds	X	F, G	
	Y		F, G
	Z		F, G, H

Figure 1.1 A schematic example indicating the projectibility of predicates F, G, and H for several kinds. Z is 'more natural' than Y, and Y is 'more natural' than X.

all line up along one dimension, and so naturalness does not form a continuum. He distinguishes two dimensions of variation:

> The value of a lawlike generalization can vary along two independent dimensions, which we might call *scope* and *force*. *Force* is a measure of the reliability of predictions made using that generalization. *Scope* is a measure of the size of the domain over which the generalization is applicable. A theoretical category about which there are generalizations of considerable scope and force is more natural than one about which generalizations tend to have more restricted scope and lesser force. (1999, p. 217)

I constructed the schematic examples so that these criteria did not come into conflict. X had the same scope but less force than Y, and Y had the same force but a narrower scope than Z. We can easily imagine less congenial examples, like a kind Z^* for which $F, G, ..., L$, and M are weakly projectible. Z would have the advantage in force, but Z^* would have the advantage in scope. As Griffiths admits, 'There will not always be a clear winner when we compare two sets of theoretical categories on the basis of scope and force' (1999, p. 217). As with the schematic case of Z and Z^*, scope and force may pull in different directions.

Comparability can also fail in another way: Different kinds might have incomparable scopes. In all of the schematic examples so far, I constructed cases which had strictly wider scopes than the previous ones. Consider instead $Z^\dagger$ for which G, H, and I are strongly projectible; again F, G, and H are projectible for Z. $Z^\dagger$ and Z have some projectible predicates in common, but each also has some that the other does not have (see Figure 1.2). As Griffiths puts it, 'The scope of generalizations made with one set of categories may overlap rather than include the scope of generalizations made with the other taxonomy so that neither taxonomy can be discarded without loss of understanding' (1999, p. 217).

If natural kinds form a hierarchy, as they do in Aristotle's metaphysical picture, then cases like $Z^\dagger$ and Z are ruled out in principle. So one might try to shore up the induction assumption by way of the hierarchy assumption. This point merits separate discussion (see § B.9, below). Let's set it aside for now.

In our schematic case, $Z^\dagger$ and Z have the same number of projectible predicates. So it might be tempting to say that they are similarly

	Predicates	
	weakly projectible	strongly projectible
Kinds	Z	F, G, H
	Z* F, G, H, I,... M	
	Z†	G, H, I

Figure 1.2 A schematic example indicating the projectibility of some predicates for several kinds. Z has greater force but less scope than Z*. Z and Z† have the same force for overlapping sets of predicates. There is no straightforward way to call any one of these 'more natural' than the others.

natural. This would escape an aspect of incomparability by collapsing scope onto a single number and ordering kinds by the raw size of their scope. Although this is easy enough in the schematic cases, this is only because the predicates are specified as a countable list. Actual cases are harder. For example, consider that ducks are typically alike. We might construe this as a small number of properties (duck-like appearance and physique) or as a great many (enough details of feathers and physique to fill an almanac). To take another example, a physical object might interact with electromagnetic waves of many frequencies across the spectrum. If we include visible color among the properties under consideration, however, we count reflectance of visible light for more than interaction with radio waves.

Even if there were some objective way of tallying up properties, scope and force are not independent. We can modify our example to illustrate this point. Suppose that the property G is less strongly projectible for the kind Z than the property H is, but that H is less strongly projectible for Z† than G is. How should we weight this in counting up the number of projectible predicates for each? It might be tempting to say that a predicate that is perfectly projectible should be counted fully and that a predicate that is only imperfectly associated with a kind should be weighted by the proportion of the kind to which it applies. In our example, if 95 per cent of Zs were G and 99 per cent of Zs were H, then G would count .95 toward Z's measure of naturalness and H would count .99. Yet the fact that the projection would fail in 5 per cent or 1 per cent of cases might not be all that is important. It might matter *which* members of Z fail to be Gs. The connection might be entirely stochastic, in which case there would be nothing further to say, or there might some systematic connection between cases that are *G*s and ones that are *H*s. The ranges of instances across which projection succeeds might themselves be complicated and overlapping, rather than nested one inside the other. So force comparisons become hydra-headed just in the way that scope comparisons did.

There is also something too quick about thinking that there is a determinate proportion of the kind for which some predicate is projectible, as we assumed in stipulating that 95 per cent of *Z*s were *G*. There will be a determinate value if we interpret this to mean the proportion in the population of presently existing *Z*s, but that is not obviously the number that matters. We might include past or even future *Z*s, but this might still not capture what we want. The problem here is similar to familiar difficulties with frequency interpretations of probability. Just as a fair coin will probably not turn up heads in *exactly* half of any finite series of flips, a kind *Z* might not reveal its true connection to *G*ness in any finite collection *Z*s. The problem is even worse, because members of a kind will not necessarily be as well defined as coin flips. There is a determinate kind of setup that counts as a flip, so we can think of them as independent and identically distributed (IID). A natural kind might have subgroups that make a difference, so the members are not IID. This undoes Griffiths' suggestion that a 'kind is (minimally) natural if it is possible to make better than chance predictions about the properties of its instances' (1999, p. 216). There is no default probability measure according to which one can make chance predictions. There is only a natural measure when we make assumptions about the kinds involved, so we cannot use that measure as an independent constraint on what will count as a natural kind.

One might escape this muddle by appealing to Mill's original notion of kinds. Recall that Mill introduced 'Kind' as technical jargon and that later commentators dubbed this his doctrine of *natural kinds*. White things do not form a real kind, because they are only similar with respect to their color and other qualities that are a consequence of color. In contrast, according to Mill, 'a real Kind ... is distinguished from all other classes by an indeterminate multitude of properties not derivable from one another ...' (1874, bk. I, ch. vii, § 4, p. 81). The idea is that the members of a natural kind share indefinitely many features. They should form a central part of our inductive enquiry, because there are always further regularities about them to be discovered. Mill counted **animals, plants, sulphur,** and **phosphorus** as real kinds because 'a hundred generations have not exhausted the common properties of animals or of plants, of sulphur or of phosphorus; nor do we suppose them to be exhaustible, but proceed to new observations and experiments, in the full confidence of discovering new properties which were by no means implied in those we previously knew' (1874, bk. I, ch. vii, § 4, p. 81).

If we accept this suggestion, then none of the schematic examples above would count as natural kinds. For each, I specified a finite list of predicates to be projectible. The toy examples, as such things often are, turn out to be illustrative rather than robust.

Regardless, Mill's suggestion fails for actual scientific examples. Recall Quine's idea that scientists work to replace observable similarity with a more refined theoretical similarity. Even if similarity is not expunged entirely, it becomes more refined. Charles Sanders Peirce objects to Mill on such grounds, complaining that 'the man of science is bent upon ultimately thus accounting for each and every property that he studies' (Baldwin 1901, p. 601). Consider phosphorus, which we now know to be the chemical element with atomic number 15; each atom of phosphorus has fifteen protons in its nucleus. Differences between the highly-reactive white and the more-stable red phosphorus are understood in terms of the arrangement of phosphorus atoms. In either case, atoms are arranged in groups of four. In white phosphorus, these tetrahedra are arranged one way; in red phosphorus, they are arranged another way. There are no doubt further properties of phosphorus which are not *derivable* from microstructural features like these, but this seems like an epistemic rather than a metaphysical point. The microstructure is responsible for indefinitely many macroscopic similarities between samples of phosphorus, even if scientists can only work out some of the details.

If we followed through with Mill's criterion, then finding out that the indefinite catalog of features could be explained by a specified core of features would show that **phosphorus** was not a natural kind after all. This seems ridiculous, since we think of chemical elements as some of the canonical examples of natural kinds. Discovering that all phosphorus shares microstructural features does not exclude **phosphorus** from being a natural kind – quite the opposite, the discovery reassures us that it is a unified kind, rather than something we had just hobbled together. (As an aside: This is not a point about reference. The discovery of the chemical structure of phosphorus might but need not be taken to show that the word 'phosphorus' has necessarily always referred to stuff with that structure. See § B.7.)

We should not expect every case to work out as nicely as the chemistry of phosphorus, but the single counterexample is sufficient to undo Mill's approach.

Taking stock of the induction assumption

The induction assumption is historically central to the tradition of theorizing about natural kinds, so it is reasonable to think of it as the canonical characterization. We should require any account of natural kinds to reckon with it. Yet we have seen that the assumption is incomplete and unsatisfactory in two important respects. First, it suffers from similarity fetishism. Second, it allows naturalness to become a slippery, unmanageable matter of indefinitely many degrees. So it does not suffice, on its own, to define natural kinds.

B.2 The essence assumption

It is often presumed that natural kinds must have *essences*, but I will not be using the word 'essence' much. Importantly, the natural kinds tradition did not begin by associating them with essences. As we saw in the previous section, Mill's conception of real kinds is rather more empirical. Philosophers who are concerned with essences – either to embrace them or reject them – tend to portray the past as more concerned with essences than it actually was. Karl Popper, for example, railed against a history of 'methodological essentialism' according to which 'it is the task of ... "science" to discover and to describe the true nature of things, i.e. their hidden reality or essence' (1966, p. 31).

There is a standard story about the history of biology which portrays pre-Darwinian biologists conceiving of species as metaphysical essences and Darwin dispelling the Hellenic-scholastic darkness. This story is, in many ways, a myth (McOuat 2009; Wilkins 2009). There is perhaps some sense of 'essence' in which pre-Darwinian thinkers thought that species had essences – but there are many and different senses of 'essence'. The word is crypto-omnivorous jargon. In the Lockean tradition, a kind is said to have a 'real essence' just if there are some facts about the world that correspond to the unity of the kind. Of course, anyone who believes in natural kinds thinks they have essences in this sense. Even though we might disagree about which features of the world correspond to them, natural kinds are more than mere names.

Another usage identifies essences with sets of necessary and sufficient conditions for kind membership. If these conditions must be perfectly precise, then this just amounts to the assumption that natural kinds must have sharp boundaries; I deal with this in § B.6, below. If the conditions may be vague, then every kind would have an essence just due to the banal fact that any kind admits of some description.

Yet another usage identifies essences with properties which are held together by laws of nature; regarding the relation between natural kinds and natural laws see § B.4, below.

Another assumption associated with essentialism is that the defining character of a kind must be *intrinsic* rather than relational; see § B.8, below.

Other usages describe essences ontologically, as abstract objects which exist over and above the entities which constitute the kind. The clearest of these suggestions is that essences are universals, but there are more obscure possibilities. Importantly, even paeans for metaphysics allow that natural kinds can support induction, should figure in science, and so on. I will focus on those clearer assumptions. The question of whether the categories that fulfill them have some special metaphysical luster is a further question. It is not a question about whether there *are* natural kinds, I suggest, but rather about the deep metaphysics of natural kinds. My project here is not to settle ages-old debates about the ontic substratum of general categories. Nevertheless, I will revisit the relation between natural kinds and metaphysics in Chapter 4.

In conclusion, there are many different issues obscured by talk of *essences*. The issues, insofar as they bear on natural kinds, will arise under different banners elsewhere.

B.3 The science assumption

It is almost universally supposed that natural kinds are the appropriate categories for scientific enquiry. Call this the *science assumption*.

Conceivably, one could satisfy the science assumption by first giving an independent, metaphysical account of natural kinds in terms of universals, real essences, or modally robust categories. Then one could require that science be enquiry into those things. Thus, the account of natural kinds would serve as a constraint on what would *count as* science. This approach would be *a priori* philosophizing of the worst kind, and I do not see how it could possibly succeed. Perhaps the account would leave us with something that we could recognize as science, but that would just be a lucky coincidence for the autonomous metaphysical picture. It could easily be the case that 'science' constrained to fit our metaphysical prejudices would look nothing like what we would ordinarily dignify with the name.

I suggest instead that we begin with a kind of modest naturalism. Science, we think, is a pretty good way of learning about the world. It is not perfect, and there may be systematic problems or limitations. Some parts of our science might ultimately need to be abandoned, and most or all of it will need to be revised as we learn more about the world. This means that we should be fallibilists, but not that we should be skeptics. It is a constraint on our account of natural kinds, then, that they form components of successful scientific taxonomy.

One might worry that this approach is anti-metaphysical. Insofar as science is something *we* do, the natural kinds construed in this way will be things that depend on *us*. If that kind of consideration convinces you that science fails to tell us about the world, however, your problem is not about natural kinds. You do not even share the modest, fallibilist but non-skeptical view of science that I take as a starting point. In a recent volume, the editors say of Richard Boyd's account of natural kinds that '[his] realism [must be] understood in (by many metaphysicians' standards) a somewhat deflationary sense. He is a realist about natural kinds in exactly, and only, the sense in which a philosopher of science might be a realist about molecules or species or natural selection' (Beebee and Sabbarton-Leary 2010b, p. 23). For my part, I do not see how we could ask for more than this. The natural kind **molecule**, because it is a category of things, will not exist in exactly the same sense that particular molecules exist. Yet I do not see why one should expect the kind **molecule** to be more certain or more real than the particular molecules. They are both part of our best scientific account of the world. (In Chapter 4, I call this *equity realism*.)

In any case, the science assumption as I have construed it is shared even by metaphysically inflationist thinkers. Although they would not put it in terms of modest naturalism, as I have, strident advocates of essentialism acknowledge that science constrains the account of natural kinds. T. E. Wilkerson writes that 'if we are to produce an interesting account of natural kinds, we should insist that members of natural kinds, and the corresponding real essences, must lend themselves to scientific investigation' (1995, p. 31). Note that he does not intend for the essentialist metaphysics to dictate terms to the science. I disagree with the details of Wilkerson's account (see § B.8, below), but the science assumption is common ground between us.

B.4 The law assumption

It is common to suppose that natural kinds are the categories that appear in the true laws of nature. Physical laws describe the behavior of massive bodies, for example, so **massive body** is a natural kind. Let's call this assumption, that natural kinds are the relata in laws of nature, the *law assumption*.

On one account, laws are necessary relations among universals. In schematic form, 'All Fs are Gs' is a law because the property of Fness necessitates Gness. Following Stathis Psillos (2002), we can call this the Armstrong–Dretske–Tooley (ADT) view. The ADT account would solve our problem rather readily: Natural kinds could be equated with the universals.

Yet we can motivate the connection between laws of nature and natural kinds even from a metaphysically more austere standpoint. A traditional empiricist view is that laws are regularities. Accounts differ as to how to sort the law-like regularities from the accidental ones, but those details need not concern us.

The important thing is that regularity accounts and the ADT account agree that laws correspond to exceptionless generalizations. The disagreement is just about whether the law amounts to anything *more* than a particular kind of regularity. So they agree that the kind picked out in the antecedent of a law will support inductions; if 'All Fs are Gs' is a law, then G is projectible for the kind F. They can also agree that the business of science is to discover the true laws of nature. Thus, for both accounts, the law assumption fits with and reinforces the induction and science assumptions.

Both accounts also share an assumption of what Nancy Cartwright (1999) calls *fundamentalism* about laws. That is, they suppose that genuine laws must be true always and everywhere. For the fundamentalist, laws must be exceptionless. Yet all of the so-called laws in actual science are true only *ceteris paribus*. One may retain fundamentalism only at the expense of current science and insist that our science has not identified any of the actual laws of nature. This point comes out in an odd way in Lawrence Sklar's discussion of Cartwright. Sklar defends fundamentalism but adds nonchalantly that the system of laws we actually have 'is replete with enough problematic aspects that there seems quite good reason to think that it is far from the final word on what the world is like' (2003, p. 440). If laws must be exceptionless, then we do not know any of the *actual* laws. So, given the law assumption, we must say that our science has not identified any natural kinds.

If the fundamentalists are right about metaphysics, then maybe we will one day discover the true laws of nature – and we would of course want the terms in those laws to be among the terms in our science. Yet, I still want to know what the terms in our science ought to be in the meantime. I have no interest in a philosophy of science which can only say what science could look like at the end of time. I want to know: What categories are appropriate for science underway, before the true laws are discovered? This question should be of concern even to the fundamentalist. Even if there are true exceptionless laws, there is no assurance that we will *ever* be able to discover them.

A fundamentalist might say that we have discovered natural kinds if the categories of our present science are those that appear in true laws of nature, even though we do not presently know the laws. It is not clear whether this obtains, though, and the science assumption would just be satisfied (if it was satisfied) as a matter of luck.

One might insist that fundamentalism is true because of the meaning of the term 'law of nature'. Even if that were plausible as an account of laws, it is not relevant to natural kinds unless we make the further supposition that the law assumption is definitional for 'natural kind'. Natural kinds, so defined, would be the abstruse promise of a hoped-for future science. We would still want a term for the categories apt to actual science. I keep the words 'natural kinds' for the latter, rather than the former.

Alternatively, one might give up fundamentalism about laws. In thinking about natural kinds, Goodman uses the notion of *law* rather loosely. As one example, he considers the law that 'All butter melts at 150° F' (1947, p. 123). It is implausible to think of this as a necessary relation among *butterhood* and *melts at 150°ness*, but it is not merely an accidental regularity. We can rely on butter's melting – not at 150° necessarily, but at somesuch temperature. Otherwise, as my officemate quips, how could we make cookies? (Douglas 2010). In Goodman's idiom, this is just to say that approximate melting behavior is a projectible feature of butter. Whatever contribution this makes to butter's being a natural kind is already captured by the induction assumption, so putting the same point in terms of the law assumption really would not add anything. This point generalizes. Laws can support inductions, and law-like generalizations can be seen as a summary of permissible inductions.

To conclude, the law assumption is subject to a dilemma. Either fundamentalism is correct or it is not. If fundamentalism is incorrect, then laws are just descriptions of areas that support inductive inference. So there is nothing lost in keeping the induction assumption but giving laws a pass. Contrariwise, if fundamentalism is correct, then the connection to natural kinds becomes a substantive hypothesis. It depends on further metaphysical and epistemological conditions. I am not concerned here either to attack fundamentalism about laws, to defend it, or to immunize it through linguistic stipulation. I merely want to say that we should not shackle our conception of natural kinds to it.

B.5 Artifacts and artificial kinds

It may be tempting to distinguish, as psychologists do, between natural kinds and artifact categories. In ordinary usage, 'artificial' is an antonym for one sense of 'natural'. Marking this distinction allows psychologists to explore the extent to which subjects categorize natural objects (like zebras) differently than they do artifacts (like automobiles).

Susan Gelman employs psychologists' standard usage when she writes, 'An object is natural if it was not constructed by humans' and natural kinds are 'basic-level categories of naturally occurring objects' (1988, p. 69). For Gelman, the distinction is closely tied to which categories are appropriate for induction. She explains:

> At minimum, a category that promotes induction is structured so that category members are similar to one another on certain

important dimensions. Objects that appear to be very similar in their known properties (thus forming 'homogenous' categories) may often share underlying similarities as well. For example, zebras apparently share many obvious similarities; appropriately, the class of zebras promotes many novel inductions. (1988, pp. 67–8)

Of course, we also draw inductive conclusions about automobiles. Yet, because we make cars, we could make them *differently*. Gelman writes, 'Artifacts are inherently varied ... and knowledge of this variation must comprise much of the expertise in the field. It may be that artifact experts learn complex sets of distinctions, whereas natural kind experts learn more general laws describing entire categories' (1988, p. 69). So, for Gelman at least, this notion that natural kinds are distinct from artifact kinds quickly becomes bound up with assumptions that natural kinds support inductive inferences and are related by general laws.

Whatever use it might be for psychologists, I argue that we should not accept the psychologists' distinction as a constraint on our philosophical account of natural kinds. Consider a succession of cases.

First, an artifact category like **automobile** supports a great many inductions. This is clear from our automotive practices. When I traded in my 1988 Volkswagen Jetta for a 2003 Toyota Echo, I was able to test drive the Toyota without difficulty. Neither of these cars was the model or year I had driven when learning to drive, but my skill at driving tolerably well generalizes. Of course, there are some relevant distinctions to be made among automobiles; being able to drive an automatic-transmission car does not transfer without practice to driving a stick-shift, for example. Yet even that difference is irrelevant for my relation to *other* cars that share the road with me. I can rely on regularities in their behavior. If I could not, then I would be unable to drive in even light traffic. Other cars and other drivers are not a perfectly homogeneous category, but they are reliable enough that I can drive from place to place with only rare accidents. The differences between cars may be important for other purposes, of course, but there are no doubt also differences between zebras.

Second, specific models of automobile support even more inductions than **automobile** in general. Ruth Garrett Millikan gives the example **1969 Plymouth Valiant**. All of them came from the factory built to certain specifications and, 'other things remaining stable, '69 Valiants ... are things that tend to persist, maintaining the same properties over time ...' (Millikan 1984, p. 279). She was writing in 1983, and there are certainly fewer '69 Valiants now, decades later. As she notes, however, the decline of the aging Valiants is something that can be predicted. Such a car 'tends to have certain properties if it has been subjected to certain conditions. ... For example, the fenders of a '69 Valiant that has not been garaged tend to rust out whereas the body stands up much better ...' (Millikan 1984, p. 280). Such facts would allow us to extrapolate from one '69 Valiant in a junkyard to what others might be like.

Third, consider **dogs**. Dogs admit of considerable variation – chihuahuas to mastiffs – and the great variation among dogs is the result of deliberate and artificial selection. A craftsman did not make Lassie out of component meat and fur, but breeders did craft collies as a breed. As Jerry Coyne quips, 'If somehow the recognized breeds existed only as fossils, paleontologists would consider them not one species but many' (2009, p. 126). Nonetheless, for reasons of systematic biology, we count the kind as the subspecies *Canis lupus familiaris*, part of a species that also includes wolves. Such decisions are based on considerations of the kind's structure or relation to other kinds. The fact that much of that structure is the result of human effort is irrelevant.

Fourth, consider a transuranic element like **Darmstadtium** (atomic number 110). It was first produced in 1995, and even now fewer than a hundred atoms of it have ever been produced. Most isotopes decay in less than a second. The longest-lived (^{281}Ds) has a half-life of 1.1 minutes. One might imagine an exotic place somewhere in the universe where atomic phenomena regularly produce Darmstadtium, so that it is not only naturally occurring but is also in some sense persistent, but that place would be nothing like Earth. All of our Darmstadtium here is the result of human artifice. Yet its structure, production, and decay are all systematically tied to physical considerations. We do not hesitate to think of it as a natural kind.

Still, there is a sense in which natural kinds should not be the result of artifice. Particular objects should not be grouped together in a way which is merely the whim of enquirers. Although psychologists look to whether the *objects* are made by us or not, what matters for natural kinds is not the objects but the *group*. This is what Carl

Hempel calls 'the familiar vague distinction between "natural" and "artificial" classifications' (1961, p. 14, 1965, p. 146). If our practical and theoretical endeavors would do as well with one taxonomy as any other, then we should not say that one is more natural than another. The suggestion that natural kinds must depend on the world in some way, rather than *merely* depending on us, is surely correct. But that is not the distinction marked by psychologists.

B.6 The sharpness assumption

It is sometimes assumed that natural kinds must be *categorically distinct*, meaning that they must have sharp boundaries. For any individual, it must be clearly a member of the kind or not. As we have seen, this is not a requirement that follows from the induction assumption. Just as predicates may be projectible with differing degrees of strength and scope, the kinds which sustain the projection may also vary in distinctness. It does not follow from the science assumption either. Although some scientific categories may be crisply defined, they cannot all be. All that is required for science is that there be clear cases which are members of the kind and clear cases which are not. Vague boundary cases only pose a problem if they are the very individuals under study.

Brian Ellis includes sharpness among the definitional criteria for natural kinds. He offers the following argument for the requirement. Natural kinds, he insists, must 'exist *as kinds* independently of our conventions' (2001, p. 19). If independence is understood too strongly, then we would have the distinction between natural kinds and artifact categories that I argued against in the previous section. Ellis would count **Darmstadtium** as a natural kind, however, so we can understand his insistence in the weaker sense I endorsed at the end of the last section. (Ellis would accept the conclusion that **dogs** are not a natural kind, but for different reasons; see § B.8, below.) So Ellis insists correctly that a kind cannot be natural if it depends merely on the people identifying it. With this constraint in place, he argues:

> Hence, where we are dealing with natural kinds, there cannot be any gradual merging of one kind into another, so that it becomes indeterminate to which kind a thing belongs. For if there were any such merging, we should have to draw a line somewhere if

we wished to make a distinction. But if *we* have to draw a line anywhere, then it becomes *our* distinction, not nature's. Natural kinds must be *ontologically* distinguishable from each other. (2001, pp. 19–20)

His argument is indirect and goes like this: Suppose, for purpose of *reductio*, that there is a natural kind which does not have sharp boundaries. Call that kind **V**. We come along and use the word 'V' to divide things into the Vs and the non-Vs. Since **V** does not have sharp boundaries, there will be some things we categorize as Vs that are not determinately members of **V**, some things we categorize non-Vs that are not determinately non-members of **V**, or both. For those objects, counting as a *V* or a non-*V* is simply a matter of how we have decided to use words. The difference depends merely on us, the people doing the categorizing, so it cannot correspond to a natural kind. This contradicts the premise that **V** is a natural kind. So it follows that natural kinds *must* have sharp boundaries.

This argument is valid, perhaps, but unsound. The contradiction results because the predicate *V* is more sharply defined than the kind **V**. Yet I see no reason why the predicate must be so precise. Surely, it is not a requirement of language use or scientific practice. Most, if not all, of the predicates that we actually use admit of vagueness and unclear boundary cases. If there were natural kinds that lacked sharp boundaries, then the correct account of them would not ascribe sharp boundaries to them. Where the boundaries are indeterminate, our language *should* be vague. Natural kinds must be ontologically distinguishable in the minimal sense that there really is some difference between the clear members of the kind and the clear non-members, but that does not require that everything that exists be clearly sortable into one of those two groups.

A similar point is made by Richard Boyd, who argues that 'a consistently developed scientific realism *predicts* indeterminacy in extension for those natural kind or property terms which refer to complex ... phenomena' (1989, p. 18). He adds further, 'What is important is that an appropriate description of the relevant facts regarding indeterminate or "borderline" cases of ... kinds ... consists not in the introduction of artificial precision in the definitions of such kinds but rather in a detailed description of the ways in which the indeterminate cases are like and unlike typical members of the kind ...' (1989, pp. 18–19). (The part that I have elided with ellipses refers to Boyd's specific account of natural kinds as *homeostatic property clusters*. I will discuss this specifically in Chapter 6, but it is inessential to the point here.)

Writing about species, T. E. Wilkerson offers a different argument which might be taken as a reason to assume sharpness. If different kinds are 'arranged in a continuum', he argues, then it is impossible that each kind 'is determined by a peculiar real essence' and so impossible that each is natural kind (1993, p. 9). However, it is not clear that this follows. There might be naturally important boundaries along a continuum, just as integer boundaries are mathematically important divisions in the real number line.

It is important to note that the assumption of sharpness is *stronger* than the assumption that there must be necessary and sufficient conditions for membership in a kind. Consider the example of colloids. (Helen Beebee and Nigel Sabbarton-Leary (2010*a*, pp. 168–70) deploy this example against Ellis, although their use of it is somewhat different.) A colloid consists of a continuous medium with small particles of another material distributed throughout. The particles are small enough that Brownian motion will keep the colloid from separating on its own, but large enough that the particles (if solid) can be filtered out. Even if these criteria are necessary and sufficient for membership in the kind colloid, they fail to define perfectly sharp boundaries for the kind. The reason is that the size of particles thus specified does not have a perfectly sharp boundary. The International Union of Pure and Applied Chemistry (IUPAC) gives a definition for 'colloidal' which specifies, in part, that the particles must 'have at least in one direction a dimension roughly between 1 nanometer and 1 micrometer' or that 'discontinuities are found at distances of that order' (IUPAC 1997). If the particles are smaller than *roughly* a nanometer – if they are not at least *on the order* of a nanometer – then the system will be a solution rather than a colloid. The IUPAC declines to draw a sharp line where there is not one in nature. Yet there is a difference in nature between colloids and solutions; blood, paint, and mayonnaise are all colloids. Ellis' argument gives us no reason to think that colloid does not form a natural – although not sharply boundaried – kind.

We have no good reason to accept the sharpness assumption, and its conflict with actual science gives us good reason to reject it.

B.7 Starting with language

The program of analytic philosophy, as a philosophical movement, was to approach any *X* by way of an examination of '*X*' and *X* terms. However, approaching every problem by way of language unnecessarily embrangled analytic philosophy in problems of meaning and translation. All philosophy turned on contentious issues in the philosophy of language. So I resist the general policy.

Be that as it may, the motivation for thinking about natural kind *terms* also comes from specific developments: the Kripke–Putnam theory of reference. Realists and anti-realists, essentialists and anti-essentialists, all feel obligated to engage the Kripke–Putnam view. Ian Hacking (2007c) provides a wealth of references to such discussions, but also argues that the so-called Kripke–Putnam view is an artifact of the subsequent literature, that Kripke's (1972) view and Putnam's (1975b) were importantly different; see also Hendry (2010). We can set the subtleties aside, because what concerns us here is the common assumption about natural kinds – and the common understanding conflates Kripke and Putnam.

The Kripke–Putnam account of reference applies at first to proper names. A person is (perhaps literally) baptized with a proper name. For example, Hilary Putnam was given the name 'Hilary Putnam'. Our use of that name does not work on the basis of some descriptions which we might associate with the name; for example, that Hilary Putnam was a central figure in analytic philosophy. Rather, the name as we use it serves to pick out the individual person Hilary Putnam in all possible worlds. This event whereby a name is conferred on a person is called a *baptism* or *dubbing*, even if there is no formal ceremony – 'baptism' and 'dubbing' become technical notions in this theory of reference. To make it more vivid, you can imagine that the baptism hooks one end of a metaphysical tether onto the name and the other end onto the person. The tether keeps the two connected even if subsequent users of the name come to believe all sorts of false things about the person. It is not broken even if you believe (wrongly) that Hilary Putnam was the Austrian logician who proved the incompleteness theorem, even if you have no other beliefs at all about Putnam. Kripke's language for this is that the name is a *rigid designator* and that the name *rigidly designates* the person.

The idea that Kripke and Putnam both had is that natural kind terms are rigid designators and work like proper names. The stock example is the term 'water'. We can imagine its initial introduction in a cartoony way: Someone held water aloft in their cupped hands and said, 'Lo, we shall call this "water".' Just as *that infant* was given the name 'Hilary Putnam', *that kind of stuff* was given the name 'water'. That kind of stuff, the fluid that was held aloft, was H_2O. Since the word picks out that kind of stuff in all possible worlds, 'water' designates H_2O in all possible worlds. This means that it is metaphysically necessary (true in all possible worlds) that 'Water is H_2O'. Of course, people did not know that water was H_2O for many centuries. It was an *a posteriori* discovery rather than an *a priori* truth. Nevertheless (according to the Kripke–Putnam story) it is a necessary truth that water is H_2O. Water's being a natural kind is supposed to consist in the word 'water' rigidly designating a kind in this way.

One way that Putnam argues for this picture is by offering his notorious Twin Earth thought experiment. We are asked to imagine that there is another distant planet which is very much like Earth, but which differs in this one respect: The fluid that flows in rivers, falls from the sky as rain, sates thirst, and so on is *not* made up of H_2O. Rather, the fluid on Twin Earth is made of different stuff and has the chemical formula XYZ. When the word 'water' was introduced on Twin Earth, someone held aloft some XYZ and dubbed *that kind of stuff* as 'water'. So *our* word 'water' would refer necessarily to H_2O, and the Twin Earthlings' word 'water' (which sounds the same as our word, but is in their language rather than ours) would refer necessarily to XYZ. People have different intuitions about this kind of case. Some people find the Kripke–Putnam story plausible and think that our Earth word 'water' would not apply to XYZ, even though XYZ is interchangeable for all purposes that do not involve chemical analysis. Other people think that 'water' means a clear potable liquid that flows in rivers and streams (etc.) and so that XYZ would count as water. (Avrum Stroll 1998 argues for this second response.) A third group resists the idea that there must be a determinate answer about how our words must refer in such odd possible worlds. They deny that there is any metaphysical tether that connects words to things. On this third view, there would be no prior fact of the matter as to whether our word 'water' applied to XYZ or not. It would depend on contingencies of how we came to encounter Twin Earth. (Joseph LaPorte 2004 is a recent advocate of this third response.) The thought experiment can be extended in lots of different ways, with travelers between Earth and Twin Earth noticing the difference before or after XYZ had been brought back to Earth – but intuitions become strained and divisive.

Mark Wilson (1982; 2006, pp. 34–8) gives a parallel thought experiment, inspired by a grade-B movie, in which druids on an island see an airplane for the first time. They call it a 'metal bird'. If 'bird' served as a natural kind term for them in the Kripke–Putnam way, then it had to be the case either that there was the metaphysical possibility of metal birds all along or that they changed the meaning of their word 'bird' when they called the airplane a 'metal bird'. Wilson argues that their prior language simply did not settle the question of whether there could be metal birds or not. The word's meaning simply did not reach out far enough to cover cases like the arrival of the airplane. The druids might insist that it did and that they were right to call the airplane a 'metal bird', but Wilson suggests that this would be a confusion. If they had not seen the airplane in flight, but only encountered it after it had crashed, then they might have called it a 'house'. When learning that it could fly, they might then call it a 'flying house'. The contingent facts about how they came to encounter it, and not some preexisting semantic fact, determine what words they use to describe what to them is a novelty.

The actual history of the term 'jade' is also a relevant example. The word is used to talk about two different minerals: jadeite and nephrite. Putnam seems to suppose that this was a mistake which was only discovered after the application of the word to instances had been settled; he writes, 'Although the Chinese do not recognize a difference, the term "jade" applies to two minerals ...' (1975*b*, p. 160). So he concludes that the word does not pick out a determinate *kind of stuff* and is not a natural kind term. Similarly, Alexander Bird (2010) argues that if 'jade' was so open-textured as to apply equally to different minerals, then the word never was a natural kind term. Joseph LaPorte (2004, pp. 94–100) and Ian Hacking (2007*a*) both look at the history of jade and find it to be rather more complicated. LaPorte writes, 'The history of the term "jade" relevantly resembles Putnam's story of Twin Earth. But ... *real* speakers did not evince intuitions of the sort that Putnam is committed to saying they would' (2010, p. 106).

Importantly, the warring intuitions in these cases are about how labels are applied to kinds, rather than about the kinds themselves.

In the Twin Earth case, we are told that there are two kinds of stuff: H_2O and XYZ. The dispute is whether the extension of our word 'water' includes only H_2O or includes both H_2O and XYZ. What does not get mentioned is that, since XYZ and H_2O are so nearly interchangeable and since the history of Twin Earth did not diverge from the path taken by Earth history until the chemical revolution, we would surely want some way of talking about them as one kind of stuff regardless of whether our word 'water' was the way to do it. We would want a taxonomy that included a category for H_2O, another for XYZ, and a third superordinate category that included both. All three would be natural kinds. Wrangling over the word is just a dispute over which pigeonhole of the taxonomic system should have the label 'water' pinned to it.

The case of Wilson's druids is similar. After the events depicted in the film, the druids' system should recognize **birds made of meat**, **stationary houses**, and **airplanes** as distinct kinds. The historical contingency that Wilson discusses is just about how their language will accommodate these three taxa.

The case of jade is different in this regard. Grouping nephrite and jadeite together might just be a convention born of historical contingency. Jade is not a chemically unified kind and so it fails to be a natural kind *for chemistry*. (The qualifier 'for chemistry' will become crucial in § B.11.)

B.8 The intrinsic feature assumption

It is often assumed that the criteria for membership in a natural kind must be intrinsic. Both Kripke (1972) and Putnam (1975b) give the examples of **gold** and **tiger**. They presume that the kinds will be distinguished by internal structure: atomic number (for **gold**) and genetic structure (for **tiger**).

The assumption is associated with the Aristotelian idea that the real divisions in nature are distinguished by essences. The connection is illustrated in the work of T. E. Wilkerson, a neo-Aristotelian who holds that 'the notion of a natural kind must be tied to that of a real essence. That is, whether we are talking about kinds of stuff (gold, water, cellulose) or kinds of individual (tiger, oak, stickleback), members of natural kinds have real essences, intrinsic properties that make them members of the relevant kind' (1988, p. 29). He takes this assumption to be intimately connected to what I have called the induction and science assumptions. He writes that

> natural kind predicates are inductively projectible, whereas other predicates are not. If I know that a lump of stuff is gold, or that the object in front of me is an oak, I am in a position to say what it is likely to do next, and what other things of the same kind are likely to do. I know for example that the gold cannot turn into water, and that the oak will not in due course produce tomatoes. ... Certain outcomes are ruled in, and others are ruled out, by the real essences of gold or oaks. (1988, p. 31)

There are several reasons for resisting the assumption that natural kinds must be intrinsic. (Regarding the connection to induction, see also Chapter 5 § A.)

First, there are formal tricks for cooking up instrinsic properties. A two-place relation over a domain of n objects can be replaced with n different one-place predicates. In this way, a relational feature can be reconceived as really being a great many intrinsic features. There is a violation of parsimony, perhaps, but the cost of adding these many different properties must be balanced against the benefit of having intrinsic properties rather than relations. This trick seems to be at work in essentialist thinking. Its essence is supposed to indicate what a lump of gold 'is likely to do next', but the lump's solubility in *aqua regia* is not something it exhibits sitting safely on the table. Rather, it is a *disposition*. Although it is tempting to think of the disposition as a relation between this lump and *aqua regia* of the form 'x will dissolve in y', the formal trick elides the second term to yield a property 'x will dissolve in *aqua regia*' which holds of the lump alone. The disposition can be seen as intrinsic to gold in this way, but the relation to the solvent is only swept under the rug.

Second, the focus on intrinsic properties and the corresponding formal fixation on monadic predicates looks like a hangover from Aristotelian logic. It is complicit in the fixation on induction as straight-rule projection, inference from 'A is F' to 'All As are F'. It is also complicit in a constrained notion of laws, where the standard schematic example of a law is 'All Fs are Gs'. This was excusable for Aristotle himself, since the best logical system available – his own – was just a class logic. We have had more powerful logical systems for more than a century now, and it is time to get over it. Relations are real.

Third, the insistence on intrinsic properties makes a terrible hash of the science. In the passage quoted above, Wilkerson (1988) gives numerous examples of biological kinds: **tiger**, **oak**, and **stickleback**. John Dupré calls him to task for this, pointing out that these examples misfire: 'The various species [of oak] have similarities and differences. The Wilkersonian inductivist who, on the basis of observation of, say *Quercus robur* (a British oak), spent an autumn waiting for the leaves to fall from, say a Cork oak or Holm oak, would be frustrated, since the latter are both evergreen' (1989, p. 249). Wilkerson (1993, 1995) takes this as a challenge and considers in more detail how to construe biological categories. So we turn to Wilkerson's reply.

That way lies madness

If a biological kind is to be characterized by some intrinsic feature, one might look to traditional morphological criteria. However, the attempt to define species in terms of surface similarity ultimately fails. There is no list of surface features that an organism must possess to be a member of a species which is sufficient to distinguish it from other, similar species. So Wilkerson tries to distinguish species by genetic makeup, as Kripke and Putnam passingly do for **tiger**. Still, every species admits of a great deal of variation. Even the number of chromosomes is not constant for all members of a species. Wilkerson concedes that 'the more we attempt to isolate the genetic features that determine biological species, the more hopeless the task becomes' (1993, p. 8). So, he concludes, 'we should abandon the assumption that species are good examples of natural kinds' (1993, p. 15). Since Wilkerson recognizes that a species will not have an intrinsic essence, his essentialism forces him to deny that species are natural kinds.

His essentialism also forces him to say that, if the form of living things is to be explicable at all, there must be some *other* essence responsible for it. He argues that an organism's essence is defined by a fully specified genetic structure. Roughly speaking, 'there will be as many kinds as individuals' (Wilkerson 1993, p. 16). Brian Ellis follows Wilkerson in this, saying that 'for natural kinds we should require genotypic identity, not genotypic similarity. Hence, from the point of view of ontology, each individual must be supposed to define its own species, unless it just happens to have an identical twin or clone' (Ellis 2001, p. 175, fn. 8). This ontological stance bungles the science in several ways.

First, Wilkerson's kinds fail to do any work. It may well turn out that **individual organism** is a natural kind, but Wilkerson is forced to posit more than just the *category* of individual organisms. He posits a separate natural kind for each genetically distinct individual. In the typical case, each such natural kind will have exactly one member. It is not clear what scientists should do with such narrow kinds. Here is one idea, though – a kind that corresponds to a specific individual would explain the unity of the person's identity across time. So a narrow kind that groups together the bodily presence of the person at different times would help solve the philosophical or psychological puzzle of personal identity. The kinds that Wilkerson posits are ill-suited for that task, however. Identical twins have identical genes, as Ellis acknowledges in the passage I quoted above, and so there would be two members of the twins' narrow, Wilkersonian kind. Yet it would be absurd to export this result to an account of personal identity, treating the twins as just one person.

Second, a single individual might correspond to more than one Wilkersonian kind. For medieval philosophers, *chimera* was the stock example of a creature that does not exist. In contemporary biomedicine, the term refers to an organism with multiple genotypes. A human chimera can form if two fertilized eggs grow together and form a single embryo. Instead of developing into non-identical twins, as they would if they did not fuse, they become a single person with two genotypes. This might produce observable effects, such as eyes of different colors, but it need not. In the more typical case, chimeras 'are impossible to differentiate from single-genotype people by ordinary observation and seriously difficult to identify even with the best of the newest biomedical technologies' (Boklage 2006, p. 579). Some non-trivial number of people, then, are chimeras – perhaps as many as 10% of the population (Boklage 2006, pp. 587, 588). Since they are not determinately genotypically identical with themselves, human chimeras will not correspond to Wilkersonian kinds. The ontology of genetic identity fails to cut nature at its joints, because each chimeric individual would be cut into two kinds that fit together like a two-piece jigsaw puzzle to form the person.

Third, genetic criteria are simply focused on the wrong thing. Biologists do not define species in terms of genetic identity or even genetic similarity. Rather, there is an important historical component: all the members of a species must share a common history of descent. (I take up this requirement in greater detail in Chapter 6, especially § B.)

Fourth, Wilkerson's response misses the point. Dupré was not arguing that **oak** is not a natural kind. On the contrary, oaks are the genus *Quercus* – a kind but not a species. 'What is needed,' Dupré suggests, 'is a characterization of the notion of a natural kind emancipated from the wholly unhelpful and misleading metaphysics of essences' (1989, p. 250). I would put the point this way: In order to free natural kinds of bad metaphysical baggage, we must abandon the assumptions that natural kinds correspond to intrinsic features and that kind members must be indefinitely similar.

One might worry that I am relying too much on the science assumption. If we were to define natural kinds in terms of intrinsic essences, then everything I have said would only show that biological science is not interested in natural kinds. Yet, as we saw, the essentialists commit themselves to saying that natural kinds should be useful for science. So I take the points above to provide a *reductio ad absurdum* of the assumption that natural kinds correspond to intrinsic features.

Functional kinds

The preference for intrinsic features rather than relations is mirrored in the preference for *structural* rather than *functional* kinds. There are also further reasons not to say that natural kinds must be structural rather than functional.

Many so-called functional kinds support inductions, serve as categories in scientific enquiry, and figure in laws of nature. These reasons motivate Jerry Fodor (1974) to insist that functional kinds can be natural kinds. If we accept the category distinction between structural and functional kinds, then Fodor is right. As Mark Wilson (1985) argues, physics makes use of many kinds which are certainly not compositional and are not obviously structural. The kind **harmonic oscillator**, for example, has members that are composed of very different kinds of underlying stuff. However, Wilson also shows that the distinction between structural and functional properties will shift depending on how a system is described (1985, p. 235). This would mean that the distinction depends entirely on representational choices that we make; as I would put it, **functional category** itself is not a natural kind. (Wilson would certainly not put the point this way, but he does not use 'natural kind' in the way I do here; regarding Wilson's metaphysics, see Chapter 4 § A.4.)

B.9 The hierarchy assumption

A tradition going back to Aristotle insists that each individual belongs to just one maximally specific kind. There is another more generic kind that includes everything of the specific kind as well as others. That kind is included entirely in an even more generic kind. Natural kinds are nested in one another like stacking dolls. Call this the *hierarchy assumption*.

The most persistent modern embodiment of the assumption is the Linnaean system of taxonomy. An organism is specified as being a member of a *species*. That species and perhaps others are members of a *genus*. The genus in turn is part of a *family*, which is part of an *order*, which is part of a *class*, which is part of a *phylum*, which is finally part of a *kingdom*. These seven levels are insufficient, so subsequent biologists added categories like *superclass*, *subclass*, *infraclass*, *cohort*, and *tribe* – over a dozen extra levels. If the hierarchy assumption were true of all natural kinds, then the entire biological system would be in the category **living thing**, which would then have its place in the grand system.

Although the hierarchy assumption fits with certain metaphysical views, we should not take it as definitive of natural kinds. Rather, we should see it as a substantive hypothesis about the organization of the world. Taken in this way, it simply turns out to be false. Counterexamples are not hard to find.

The Linnaean system has failed as a system for organizing living things (see Chapter 3 § B). Some species are grouped according to criteria of interbreeding, others are grouped by phylogenetic or ecological criteria. There is no assurance that such criteria will not group some particular organism into two overlapping, non-nested kinds. And there are other biologically important kinds that cut a wider swatch across species boundaries. The kind **predator** is determined by organisms' place in an ecology, so one creature might be a predator even while a genetically similar individual (in a different environment, following different developmental pathways) might not.

The difference between different phases of matter is important for chemistry and parts of physics, and there are a great many generalizations that can be made about (for example) solids. So the induction and science assumptions suggest that **solid**, **liquid**, and **gas** are all natural kinds. Since solids may be constituted of different kinds of stuff, these phase kinds will not nest hierarchically into a system of compositional kinds. So a defender of the hierarchy assumption must deny that they are natural kinds. I see no reason to do so beyond the hierarchy assumption itself.

Protein and **enzyme** both seem to be natural kinds but are not nested one in the other. Emma Tobin explains, 'even though renin and the hairpin ribozyme can be classified together as *enzymes* and renin and albumin can be classified together as *proteins*, albumin and the hairpin ribozyme are not classified together as either *enzymes* or *proteins*, since not all enzymes are proteins and not all proteins are enzymes' (2010, p. 183). Howard Kahane (1969) and Muhammad Ali Khalidi (1993) provide a flurry of further examples. Even if many of these could be defused, it seems as if non-hierarchical kinds are just too rampant in science to be dismissed entirely.

So I recommend we abandon the hierarchy assumption. As a substantive hypothesis about how natural kinds are in fact organized, it turns out to be false. We can abandon it without thereby giving up the notion of a natural kind.

We saw before (in § B.9) that the hierarchy assumption would have filled some of the gaps in the induction assumption. It would give us an understanding of why kinds support the inductions that they do: More specific kinds would support stronger inductions but over a narrower range. The more generic kinds would support weaker inductions but with greater scope. This collapses once we consider inductive inferences using overlapping, non-hierarchical kinds. For example, consider a female jaguar. By observing her, we might generalize about female mammals, predators, or any number of other kinds which do not neatly fit inside one another. The range of inductions which we can make about each category varies along indefinitely many dimensions. It cannot be collapsed to a single degree of naturalness. So the induction assumption remains muddled. (The point here is not that the muddle is inescapable, but just that the hierarchy assumption does not provide an escape. I argue that the account of natural kinds which I develop in the next chapter resolves this muddle; see Chapter 2 § C.)

B.10 The scarcity assumption

It is sometimes assumed, usually implicitly, there will be only a short list of natural kinds. One might optimistically hope for the world to be constrained to a dozen categories, like Kant's table of the categories of understanding. Regardless of the number, what these ideas share is the assumption that natural kinds are sufficiently sparse in the world that they could all be cataloged in a suitably subtle theory. One might instead be resigned to an infinity of natural kinds but still hope that there is some tractable way to encode the whole infinite structure.

Sometimes this hope is underwritten by the thought that the basic ontology of the world is the smallest parts of things. If we take the list of fundamental particles from the physicists' standard model, then we get six varieties of quark, six varieties of lepton, and a short list of bosons. Even supposing that this is right as a matter of composition, however, nothing follows about natural kinds. Once we have given up the assumption that all natural kinds must be compositional, as I argued we should (§ B.8), then nothing about the number of natural kinds follows from the number of compositional building blocks.

Sometimes scarcity is seen as a requisite for realism. Hilary Putnam characterizes *metaphysical realism* as the view that 'the world consists of some fixed totality of mind-independent objects' and that there is 'exactly one true and complete description' of that totality (1981, p. 49). The argument for scarcity would then go like this: Whatever structure the world has must be expressible in the one true theory. The theory itself must be expressible in some language. So the number of natural kinds can be at most a tractable infinity.

Yet, as Richard Boyd has often emphasized (e.g. 1989, p. 20), it is unclear why a realist should accept the premise that the world is simple enough that everything about it could be recorded in some complete description. We cannot *a priori* rule out the possibility that there are more natural kinds in the world than could be captured in even a sophisticated future science. Given the actual complexity of the world, I suspect that this possibility is realized – but the point here does not depend on weighing metaphysical hunches. The point rather is that the scarcity assumption is contingent on how complicated the world actually turns out to be. We should not accept it as a constraint on our account of natural kinds.

B.11 The implicit simpliciter assumption

Many writers presume that a kind is either a natural kind or it is not, as a feature of the whole world, without any further specification required. This assumption is rarely stated explicitly by philosophers who are committed to it. As such, it lacks a common name. Let's call it the *simpliciter assumption*.

It sometimes peeks through philosophers' more grandiose claims. For example, the nineteenth-century logician Carveth Read presumes that all natural kinds must belong to one system of kinds. He writes that 'Natural Classification ... would have a place for every thing and every event in the world, according to its closest affinities, displaying the whole hierarchies of Natural Kinds and Causes' (1878, p. 248). Natural kinds, he presumes, will be natural in virtue of their connection to *everything* else. So a kind cannot be a natural kind in some respect or in some domain, but not in others.

It is also often presumed in the way that the problem is posed. Asking the question 'Is X a natural kind?' presupposes that there is a univocal yes-or-no answer – that there is a fact of the matter about the kind, that it is either a natural kind or it is not. Thinking that natural kinds have essences invites us to think of a monadic second-order property 'X has an essence' – a feature that holds of a kind in itself, not considered in relation to any other thing – and so to presume the simpliciter assumption.

Alexander Bird and Emma Tobin (2009), surveying possible views about natural kinds, consider only accounts that accord with the simpliciter assumption and do not acknowledge any alternative. The assumption is presumed by the stance that 'there are in fact natural divisions among things, so that when we attempt to make a natural classification, there is a fact of the matter as to whether that classification is indeed genuinely natural' (Bird and Tobin 2009). They contrast natural kinds with categories that 'reflect human interests', and there is a sense in which this is right. There are no unicorns in the world, and so **unicorn** is not a kind of thing in the world. A fascination with unicorns would not change that. Yet we only ever offer scientific accounts of things we care about. So all the categories that appear in our science reflect our interests. More importantly, the categories posed in a specific science reflect the focus of that enquiry. Particle physics posits kinds of particles, like **neutrino**. Geology posits kinds of rocks, like **granite**. Biology posits kinds of organisms, like **anglerfish**. The fact that physicists do not recognize **anglerfish** as a natural kind does not mean that it is not. Rather, it is not a natural kind *for physics*. Biologists, for their part, do not talk about neutrinos.

This is a way of rejecting the simpliciter assumption: Say rather that natural kinds are *natural* only relative to a specified domain of enquiry. Richard Boyd suggests that this assumption of enquiry relativity is the generally accepted view: 'It is widely recognized that the naturalness of a natural kind – its suitability for explanation and induction – is *discipline relative*' (1999a, p. 148). Obviously, the simpliciter assumption and enquiry relativity are incompatible. The prevalence, at least implicitly, of the simpliciter assumption shows that there is no concensus around enquiry relativity. Yet Boyd is correct to note that there are thinkers who acknowledge it.

Quine, who views kinds in terms of similarity, argues that similarity is relative to a domain of enquiry. He writes, 'Different similarity measures, or relative similarity notions, best suit different branches of science; for there are wasteful complications in providing for finer gradations of relative similarity than matter for the phenomena with which the particular science is concerned' (1969a, p. 137).

Mark Wilson identifies one sense of 'natural kind' to be 'the important physical traits for, e.g. prediction'. He adds that 'the "importance" of a physical property is not an absolute, object independent notion but depends essentially on the system under study' (1985, p. 237). So natural kinds in this sense are domain relative.

Boyd himself writes 'that "natural" kinds are relative to disciplines, inductive task or contexts of enquiry' (1982, p. 642). He even puts the point in language close to mine: 'Thus the fundamental notion in the theory of theoretical natural kinds is not the notion of such a kind, *simpliciter*, but instead the notion of a kind's being natural with respect to ... the role it plays in a disciplinary matrix' (1999b, p. 57).

I will argue that we should accept enquiry relativity – and so that we should reject the simpliciter assumption.

A direct argument for domain relativity

It is a truism that scientific inference requires reliance on auxiliary hypotheses or background theories. Knowing merely that a thing is a lump of gold will not necessarily allow me to infer what it will do when placed in *aqua regia*. I must know that gold dissolves in *aqua regia*. For this to be anything more than a brute and mysterious regularity – for me to relate gold's dissolving in *aqua regia* to other facts about gold – I must know a great deal of chemistry. If my systematic knowledge were about other things, such as botany or the stock market, I would not be able make any prediction about what this lump of gold would do. So the kind **gold** only supports inductions in the context of chemistry. By the induction assumption, **gold** is only a natural kind in the context of chemistry.

Boyd's argument for the enquiry relativity of natural kinds is similar. He writes:

> The kind **salt of sodium** is a natural kind just because reference to it contributes to the accommodation of the inductive and explanatory practices of chemists and others to relevant causal structures. Classifying reagents accurately as sodium salts and referring to them by the term 'sodium salt' would make no such contribution except in the context of a whole bunch of theory dependent classificatory experimental and inferential practices involving – among other things – reference to lots of other chemical kinds. (1999b, p. 57, my bold)

Being domain-specific does not make the regularities any less natural. These regularities about samples of gold and samples of sodium salt are not somehow *invented* by chemists. As Boyd notes, they are real features of the relevant causal structures.

There is no general canon of inductive inference. Even the most popular candidates must be supplied with information about the domain of enquiry. Bayesian confirmation theories, for example, only operate given prior probabilities and likelihoods. The general point has been argued effectively by John Norton (2003), who calls the requisite domain-specific knowledge *material postulates*. The knowledge might take the form of domain-specific rules of inference, *material inference principles* (Brigandt 2010). In either case, ampliative inference is only possible because of the domain-specific resources (Magnus 2008). So the induction and science assumptions will only ground natural kinds relative to a domain of enquiry.

Trying to recover the simpliciter assumption

If this argument is correct, then strictly speaking a kind is only natural relative to some domain of enquiry. If I say 'Water is natural kind', then I have said something semantically incomplete. Naturalness is a two-place relation of the form 'k is a natural kind for d', so I simply have not expressed a complete thought. In an actual conversation, of course, context may be enough to determine which enquiry is at issue. If we are talking about chemistry, then we know what I mean. The implication, if not my bare utterance, is true; water is a natural kind for the domain of chemistry. If we are talking about cryptography, however, it is not clear what I mean. Water is not a natural kind for cryptography.

One might try to come up with some general, context-independent way of filling in the relevant kind. The most obvious approach is to define 'k is a natural kind (simpliciter)' to mean 'there exists a domain of enquiry d, such that k is a natural kind for d'. This would recover a sense of natural kind for which the simpliciter assumption would hold. It is not clear what this would show, however.

First, as we saw when discussing intrinsicness (§ B.8), this kind of formal move is always available. The more fundamental sense of natural kind would still be the relational one.

Second, the derived concept would not be a very useful one. Suppose again that I claim that water is a natural kind, but in a conversation about cryptography. In the derived, simpliciter sense of natural kind, this would be true. Yet it would be true because of facts about chemistry, which are simply a *non sequitur* in our conversation.

Third, this would require quantifying over possible enquiries. Saying that there exists some domain d requires some characterization of what the universe of possible enquiries is like. Such a characterization would be problematic and perhaps impossible. In contrast, using the relative conception of natural kinds in a particular case only requires that the domain of the particular enquiry is among the possible ones. We need to say that this is one possible enquiry, but we do not need to characterize the whole range of all possible enquiries.

What's a domain of enquiry?

I have not given a precise analysis of what a *domain of enquiry* is, and trying to give an exact definition would lead us into a regress of revision and monster barring. Even so, existing scientific disciplines give us clear exemplars because the purview of any given discipline is a domain of enquiry. If scientists invent a novel but successful discipline at some point in the future, then the purview of that discipline will be a domain of enquiry, too. Still, what a domain is can be interpreted in two importantly different ways.

First, with emphasis on *domain*, a domain of enquiry could be a range of objects and phenomena. The domain of astronomy (in this sense) includes stars, planets, orbits, and so on. Whether we do astronomy or not, this domain exists in the world. Our efforts at space flight may make some small changes to the domain but in an innocuous causal way: leaving footprints on the moon, adding satellites to the population of extraterrestrial debris, and so on. The domain does not depend on our enquiry into it.

Second, with emphasis on *enquiry*, a domain of enquiry could depend on the structure of a practicing scientific discipline. The boundaries of a domain (in this sense) are drawn not just by what the enquiry addresses, but by the specific epistemic resources of working scientists. These resources include observational abilities (determined by the instruments that the community has), inferential powers (determined by the theories that the community accepts), methods, the social organization of the community, and more besides. The domain of astronomy in this second sense depends on what astronomers are like. It is historically variable, changing with what astronomers can observe and accommodate in their theories.

In the first sense, the planet Uranus was part of the domain of astronomy in Galileo's time. Although undetected, it was among the planets of our solar system. In the second sense, it was not. Uranus did not figure in the disciplinary matrix of Galilean astronomy.

The argument given above for enquiry-relativity does not distinguish between these two interpretations. A generalization which does not hold in an unrestricted way may still hold for objects in a restricted domain, and a predicate which is not fully projectible might nevertheless be projectible within that domain. Considering the specific domain, rather than the totality of everything, may make induction possible and science tractable. So a kind which is not natural simpliciter might count as natural relative to a domain of enquiry in the first sense. (I return to this point in Chapter 2 § C.)

Yet even if some part of the world has structure which can support inferences, that does not mean that scientists are positioned to exploit it. They can only do so if they have sufficient observational and inferential resources. Mindful of this, one might relativize natural kinds to domains of enquiry in the second sense.

I suggest that we provisionally adopt the former approach and take a domain of enquiry to be a range of phenomena, things, and features of things. Insofar as new methods transform a domain of enquiry, it is by directing scientists' attention to parts of the world they had not considered before. And differences in phenomena of interest can make for a big difference between sciences. For example, psychology and physiology are both concerned with people, but their domains include different features of people, different phenomena in which people participate, and different other things to which people stand in relation. Because of these differences in their domains, there is no reason to expect them to share many natural kinds.

Construing domains of enquiry in the first sense rather than the second foregrounds respects in which the natural kind structure of the world does not depend on what we presently believe or on which scientific enquiries we actually conduct. This is really just a programmatic remark and a promissory note. I do not have a compelling argument for construing domains of enquiry one way rather than the other. The examples which I discuss in later chapters are, I think, best understood in the first way. If that is so, then my whole account provides some reasons to favor it.

Be that as it may, much of what follows does not depend on the choice. If you prefer the second construal, you need not eschew my whole account. You can read 'domain' in the second sense, and much of what I argue will still go through. I do not think the account works *as well* that way, but it still works better than an account of natural kinds shackled to the simpliciter assumption. (For some explicit discussion of the difference, see Chapter 4 § A.1.)

C. Keeping score

In this chapter, I have surveyed roughly a dozen common assumptions made about natural kinds. There are three of these which, I have argued, are deep and important: the induction assumption, the science assumption, and domain relativity. The other assumptions either are about other things, are contingent on how the things turn out to be, or are confused – fates which are summarized thusly:

- The *induction assumption* (§ B.1), *science assumption* (§ B.3), and *domain relativity* (§ B.11) will structure the account of natural kinds given in the next chapter.

- The *essence assumption* (§ B.2) is obscure, and its substantive content is captured by other assumptions.
- The *law assumption* (§ B.4) depends on the account of laws, which is a separate and contested philosophical matter. Without some compelling reason to accept the connection, accounts of laws and natural kinds do not significantly constrain one another. If your accounts of laws and natural kinds coincide, then the law assumption is satisfied but uninformative. If they do not, then you should simply deny the law assumption.
- The *distinction from artifact kinds* (§ B.5) tracks something which might be important for psychology but which is different than what interests philosophers.
- The *sharpness assumption* (§ B.6) lacks independent motivation and gets the science wrong.
- The *approach by way of natural kind terms* (§ B.7) makes the outcome depend on perverse thought experiments and vexed questions in philosophy of language. Whether natural kind terms operate as rigid designators is a question for philosophical semantics that need not concern us here.
- The *intrinsic feature assumption* (§ B.8) is both a formal mistake and makes a hash of actual science.
- The *hierarchy assumption* (§ B.9) is contingent and turns out to be false.
- The *scarcity assumption* (§ B.10) is contingent. It looks to be false, although conceivably it could be true if future science develops in an unprecedented way.
- The *simpliciter assumption* (§ B.11) runs afoul of enquiry relativity.

In the next chapter, I offer an account of natural kinds informed by the three correct and relevant assumptions.

2
A Modest Definition

In the previous chapter, I argued in favor of three assumptions about natural kinds and against several others. The three principles that survived scrutiny are what I called the *induction, science,* and *domain relativity* assumptions – respectively (1) that natural kinds support inductive inferences, (2) that natural kinds are the categories suited for scientific enquiry, and (3) that a kind is natural only relative to a specific domain of enquiry. In this chapter, I sketch a conception of natural kinds that fits these constraints.

I begin by offering an initial formulation in two parts, which I'll call the success clause and the restriction clause. The restriction clause, in its initial intuitive formulation, is too strong. I consider two reformulations, a weaker version which allows naturalness to be a matter of degree (§ B) and a version which applies explicitly to kinds which have already appeared in successful science (§ D). I endorse both, although I naturally prefer the stronger, binary formulation when it applies.

A. First formulation

The intuitive idea is this: A natural kind is a category that scientists are forced to posit in order to be scientifically successful in their domain of enquiry.

Scientific success might be decomposed into several desiderata: *Inductive* or *predictive* success consists of accurately predicting the significant phenomena in the domain. This may include both projective inductions and more intricate kinds of predictions. *Explanatory* success involves giving a systematic account that makes sense of the domain. Where the domain is structured by causal interactions, identifying causes will be an important part of both explanatory and inductive success. One might add success at guiding manipulation and intervention in the world, but perhaps this is redundant; figuring out what will happen if I do something is a kind of prediction, and a causal account of how to make things happen is a kind of explanation. There are philosophical debates to be had about conditions of theoretical adequacy, but that could be a whole other book. For some examples at least, the general characterization that I have given will be enough. Where the details of success matter, the account of natural kinds is compatible with different ways of filling the lacuna.

With some sense of success in mind, we can then characterize natural kinds in this way:

> A category k is *a natural kind for domain d* if (1) k is part of a taxonomy that allows the scientific enquiry into d to achieve inductive and explanatory success, and (2) any alternative taxonomy that excluded k would not do so.

The first conjunct requires that a successful account of the world can be given using a taxonomy that involves the kind; let's call this the *success clause*. Its purpose is obvious: Science only gives us any reason to believe in a kind if the science is itself successful.

The second conjunct requires that the science would not have succeeded using a taxonomy that did not distinguish the kind; let's call this the *restriction clause*. It needs to be refined somewhat, a task I will turn to soon enough, but it will work as a first approximation. It serves two purposes.

First, the restriction clause deflects what is often called the tacking paradox or the problem of irrelevant conjunction. Without the restriction clause, we could start with some already successful science and add an irrelevant kind like **unicorn** to its taxonomy. The science would continue to be successful, and so **unicorn** would count as a natural kind. This strategy works for any kind and regardless of the domain of enquiry, so it would follow trivially that every kind is natural for every domain. The restriction clause avoids this calamity by requiring that the kind be indispensable to the taxonomy's support of scientific success. If we begin with an already successful science, unicorn would only be a legitimate addition if it allowed for some *further* success that would not be possible without it.

Second, the restriction clause keeps the account from attributing natural kinds to domains where scientific success is too easy. If it is a trivial matter to find a taxonomy that will support success in some domain, such that an indefinite number of entirely distinct taxonomies would do equally well, then the world is not really serving as a constraint on taxonomy. To take a banal example, suppose that the domain of enquiry is *how to spend a satisfying afternoon in San Diego*. There are indefinitely many, wildly different ways of carving up the space of possibilities which would lead to a satisfying afternoon and so to success in this domain. The kinds in a successful taxonomy should not count as natural kinds in such a case; because of the restriction clause, they will not.

This definition thus captures the intuition that scientists discover natural kinds when the taxonomy they employ is one that they need to employ in order make sense of the world. It does not provide any *a priori* guarantee that there will be natural kinds for any given domain of enquiry, because some domains might be sufficiently easy to describe that many category systems would do. It seems likely that some domains of enquiry will have natural kinds and others will not. It is an empirical question whether a particular domain has natural kinds and, if it does, what they are. There is no way to say without looking at the details of the enquiry. We should not have strong opinions, in advance of looking at the details, about how many of the categories in our present science are natural kinds.

B. More or less natural kinds

The restriction clause, as I formulated it above, is remarkably strict. It requires that, for a kind to be natural, there must be *no* taxonomy that could support scientific success that did not include that kind. Sam Page calls this 'individuative independence'; the view that the world is 'divided up into individual things and kinds of things that are circumscribed by boundaries that are totally independent of where we draw the lines' (2006, p. 327). The restriction clause as formulated so far requires natural kinds to be part of the *only* taxonomy that could support success, which amounts to requiring *total* independence of the kind Page describes.

Especially since success has different aspects, we can expect that there will be multiple possible taxonomies which could support some inductive and explanatory success in any given domain. With the restriction clause as written, this would mean that there simply are no natural kinds for that domain. Yet the important idea behind the account is not that there be some *unique* taxonomy that can support successful enquiry. Rather, the idea is that the world condemns a great many taxonomies to failure. Constraint from the world is what makes identifying natural kinds the discovery of structure in the world, rather than merely the imposition of a set of labels onto things that are undifferentiated in nature.

Acknowledging this, we can say that a kind is a natural to the degree that the world penalizes enquiry conducted using taxonomies that do not acknowledge the kind. We can put the point this way:

A category k is *natural for domain d* insofar as k has to be recognized in order to achieve inductive and explanatory success providing an account of d.

One might worry that this opens the door for weakening what counts as *natural* until nothing is left of the restriction clause. I think it is helpful to see this worry as shadowing issues that are familiar from the literature on the *underdetermination of theory by data*. So I will give a brief summary of those issues and then return to the worry about natural kinds.

B.1 Lessons from underdetermination

There are several different problems that travel under the name 'underdetermination'. One of them alleges that, for every theory, there are indefinitely many *empirically equivalent rival theories*. So, the argument concludes, there is no hope of picking out the true theory from its false but empirically equivalent rivals. As Larry Laudan (1990) has emphasized, arguments of this form often demand merely that the rival theories be compatible with the same data. Actual science, however, demands more than that; scientific inference requires more than a theory's just being logically consistent with observed phenomena. In actual cases, the form a theory can take and the inference from phenomena to theory are mediated by background beliefs. Auxiliary hypotheses, underwritten by background knowledge, can help eliminate many of the would-be rival theories. (In the previous chapter, I used this mediation by background knowledge to show that natural kinds are domain relative; see Chapter 1 § B.11.)

The necessity of auxiliary hypotheses leads to another problem of underdetermination: When the test of theory relies on background knowledge, any recalcitrant data can be accommodated either by revising the theory or by revising the background beliefs. This is often called *the Duhem–Quine problem*. As a general worry, it shows that scientific inference which relies on auxiliary hypotheses cannot yield absolute certainty. For fallibilists about science, this is not a problematic conclusion. We should never have expected absolute certainty, but we can nonetheless be justifiably more or less confident in different claims. We can make inferences that go beyond our immediate experience, even though every ampliative inference relies on some background beliefs about contingent features of the world. If the background assumptions are better secured than the theory we are testing, then recalcitrant evidence should be taken as a mark against the theory.

The standard countermove by underdetermination enthusiasts is to block the appeal to background knowledge by enlarging what counts as a *theory*. Rather than the choice between two specific hypotheses, the argument for underdetermination is made with respect to systems of hypotheses plus related background knowledge. When resources are found for deciding between these larger rivals, they are enlarged again. Ultimately, the underdetermination ends up being between *total sciences* or *systems the world*. Underdetermination of this kind is not a real problem, though. There is something suspicious about the very notion of a total science, and choices between them would not look anything like actual science. (For more detail on this dialectic see my 2005*a*.)

In summary, philosophical worries about underdetermination have three distinct aspects. First, the problem of empirically equivalent rivals is supposed to show that – for any theory – there are rivals that are just as well confirmed. The problem is deflected by an appeal to background knowledge strong enough, in particular instances, to decide between a theory and its rivals. Second, the Duhem–Quine problem is supposed to show that any recalcitrant data can be accommodated either by revising the theory that is being tested or by revising the background beliefs that legitimate the test. This problem is deflected by noting that some beliefs are more secure than others – not absolutely secure, but secure enough to motivate specific inferences. Third, the Duhem–Quine problem is re-posed as a problem of total sciences. Since there is no *background* behind a total science, there are no background beliefs to which one could appeal. This last is just an abstruse, hermetic problem and not like any real problem of theory choice.

B.2   The lessons applied

There is a worry about natural kinds that parallels the problem of empirically equivalent rival theories: Perhaps any taxonomy would admit of indefinitely many distinct rivals which could support scientific success equally well. The world might doom any particle physics that explains bubble chamber tracks in terms of **unicorns** while still allowing scientists a great deal of freedom in accommodating the data.

Just as the worry about empirically equivalent rival theories was answered by appealing to background theories that are sufficiently secure to justify an admittedly fallible inference, systematic constraints can help to constrain the possible rival taxonomies. Since scientific success requires providing systematic explanations, arbitrary or wild taxonomies will not necessarily allow for success. Systematic constraints arise both *between* domains and *within* domains.

First, between domains: The accounts of one domain must ultimately fit together with accounts in related domains. Where scientists are enquiring into one domain that overlaps a second, well-understood domain, they are not free to divide things up in any way whatsoever. They must incorporate or at least acknowledge elements of the taxonomy of the established science. For example, the taxonomy of molecular biology should coincide at least at the edges with the taxonomy of chemistry below and cellular biology above. Ilya Farber calls this constraint *domain integration* (Farber 2000). Science within a domain may insulate itself for a time, while it works on its own internal problems and explores the potential of its own resources. Eventually, however, the enquiry must address its relation to the sciences of neighboring domains. Richard Boyd makes a similar point, writing that 'reference to natural kinds facilitates induction and explanation with respect to a wide variety of issues – often beyond the domain of a single scientific discipline as these are ordinarily understood' (1999c, p. 81).

Second, within domains: Enquiries in different domains need not import the taxonomies of other enquiries in all their detail. An independence of taxonomy for matters central to the domain – even if only a limited independence – is an important consequence of the fact that natural kinds are relative to particular enquiries. Edouard Machery gives the example of **life**, which is understood in different ways by diverse disciplines such as evolutionary biology, molecular biology, artificial life, synthetic biology, astrobiology, and research into the origins of life. Machery treats 'life' as a term of scientific jargon and argues that 'different definitions tend to be preferred in different disciplines because these disciplines have different agendas' (2011, § 4.3). From these differences, Machery argues that **life** is not a natural kind. He writes:

> One could ... object that if **living beings** are a natural kind, all the disciplines interested in the definition of life will end up with the same definition. It is however unclear whether living beings form a single natural kind, since nature rarely yields a unique way of classifying the world. Given that the disciplines under consideration have different agendas and focus on different phenomena, it is plausible that the most useful classification of phenomena – for example, for inductive purposes – will vary across them. (2011 § 4.3)

Machery's argument here begins with what I have called the simpliciter assumption, supposing that a natural kind must be natural for all enquiries. If this were so and if different enquiries found different categories under the heading 'life', then **life** would fail to be a natural kind. As I have argued, however, we should reject the simpliciter assumption and say instead that kinds are natural kinds only relative to specified domains of enquiry (Chapter 1 § B.11). The different enquiries may, as Machery says, be forced to construe *life* in different ways in order to achieve inductive success within their separate domains. Whereas most biological disciplines (concerned with terrestrial life) need not worry about anything besides carbon-based organisms, astrobiology (concerned with the possibility of life on other planets) and artificial life (concerned with a generalized engineering domain) have broader but differing requirements. At points of engagement between these disciplines, it is important to be clear that they do not use the word 'life' in the same way. Their separate constraints mean that they might nevertheless identify distinct but equally legitimate natural kinds. So there would not be a single natural kind **life**, but instead several related kinds which figure in the work of different disciplines.

Disciplines with partly coincident domains should, where possible, provide compatible taxonomies. At the same time, systematic concerns within particular domains may force enquirers to devise their own special categories. In this way, systematic concerns provide considerable constraint on what counts as a successful taxonomy.

The move to total science (in the underdetermination debate) suggests a similar move that one might make in trying to block the argument from systematic constraints. The objection would be this: Combine together the domains of the different enquiries that Machery discusses, so as to yield one super-domain. The enquiry into this super-domain would not require a determinate category of **life**, so there is no natural kind **life**. Even if some specific categories might fit the needs of some larger domain, the domain could be enlarged further by combining it with other disparate domains. Ultimately, concatenating domains in this way generates a domain which is the whole universe in all its detail. This parallels the shift in discussions of underdetermination from thinking about theories to thinking about systems of total science.

This objection presumes that there are no limits on how domains might be combined. However, an enquiry into the total domain – the domain which includes everything – would be science *simpliciter*. The natural kinds for it would be natural kinds *simpliciter*, so concatenating domains without bound is tantamount to accepting the simpliciter assumption. It gives up on the idea that natural kinds are relative to specific enquiries. Just as the choice between total sciences is nothing like actual science and so becomes too abstruse to be taken seriously, bloated super-domains could not be the domain of any actual scientific enquiry. Cosmology plus macroeconomic behavior does not form a single domain, because enquiry into them necessarily runs along different lines. Similarly, the fact that evolutionary biology is concerned just with life as it is on our planet and astrobiology is concerned with life as it could be elsewhere makes a big difference for their respective taxonomies.

C. Induction redux

In the previous chapter, I raised two complaints about natural kinds understood simply in terms of the induction assumption. I raise them again here to argue that my account answers them. (See Chapter 1 § B.1.)

The first complaint was similarity fetishism. It resulted primarily from the narrow inductivist focus on *induction* as projective inference. For similarity fetishists like Mill, Goodman, and Quine, a kind could only be a collection of similar things – ones sharing a great many features. The notion of scientific success that appears in my conception of natural kinds is broader than just success at projective induction. Explanatory success will allow for kinds that are importantly more subtle than mere clusters of similarity. A biological species, for example, is a lineage of organisms held together by various causal forces. The details of these causal forces – both the long history of the species and the developmental histories of individual organisms – will account both for shared features and for features which systematically vary in the population. I develop this example in greater detail in Chapter 3 § B and Chapter 6.

The second complaint was that inductive power can vary with respect to strength and breadth in various directions. So the focus on projective induction made *naturalness* worse than a matter of degree – rather, it made naturalness a multifarious matter of indefinitely many degrees and respects.

One might worry that the version of the restriction clause which allows a kind to be more or less natural suffers from the same problem. However, natural kinds arise within a domain of enquiry, and the domain constrains which objects and properties matter – which are centrally important, which are peripheral, and which are not even at issue. It thus serves to constrain the range of circumstances across which a kind needs to be robust. For example, the categories of classical mechanics and chemistry allow for indefinitely many precise predictions. They break down in cases where relativistic and quantum effects come to the fore, so they will not be natural kinds for domains of enquiry that include phenomena at the atomic or cosmic scale. Yet kinds that are natural for those domains, such as **top quark**, will not allow for substantive inferences in quotidian domains. So, I suggest, the *matter of degree* left in my account is less slippery.

To make this point in a schematic case, recall the example summarized in Figure 1.2. In that case, we imagined three kinds: predicates F, G, H are strongly projectible for Z, predicates $F, G, H,..., M$ are weakly projectible for kind Z^*, and predicates G, H, I are strongly projectible for $Z^\dagger$. Considerations of induction cannot settle which of these is more natural *simpliciter*. Now suppose we are considering the matter relative to the domain of some specific enquiry. Suppose further that the predicates F, G, and H describe entities or phenomena in the domain, but that the predicates $I,..., M$ do not. For this domain, Z has clear advantages over the other two – in strength and breadth, respectively. This is illustrated in Figure 2.1. The example shows that the focus provided by a domain of enquiry might settle matters of degree which could not be settled by considerations of induction alone. The extent to which this happens in actual cases can only be revealed by considering real examples.

To sum up: The comparative version of the restriction clause suffices to provide a non-trivial sense to 'natural kind'.

But it is possible, at least in some cases, to refine the restriction clause so that kinds in a domain are natural or not as a binary matter. I pursue this strategy in the next section. Where the strategy works, we can skirt matters of degree entirely. Where it does not, the relative version of the restriction clause still allows us to ask whether a kind is imperfectly natural within a well-defined domain.

	Unrestricted Predicates		Domain-restricted Predicates	
	weakly projectible	strongly projectible	weakly projectible	strongly projectible
Kinds Z		F, G, H		F, G, H
Z^*	$F,..., M$		F, G, H	
$Z^\dagger$		G, H, I		G, H

Figure 2.1 There is no straightforward way to call any one of the kinds 'more natural' than the others *simpliciter*. For a domain which includes the phenomena or objects picked out by F, G, and H but not those picked out by the other predicates, however, Z is clearly the more natural kind.

D. Natural kinds for settled science

We can weaken the restriction clause so that a kind's naturalness is a matter of degree, and doing so does not empty natural kinds of all their philosophical interest. Nevertheless, it is worth at least trying to apply a stronger, binary notion. If we are asking about a given successful enquiry, we can state the original restriction clause as a counterfactual:

> For a successful science of domain d, the kinds that the community has identified are *natural kinds* if the community would *not* have attained success had it *not* employed a taxonomy that includes those kinds.

Like the first formulation, this avoids the tacking paradox (if the same taxonomy *minus k* would have underwritten the same success, then k is not a natural kind) and disqualifies cases in which success is too easy (if an utterly distinct taxonomy would have underwritten the same success, then none of the kinds in the taxonomy are natural kinds).

In the next section, I consider some familiar issues from the history of chemistry and use this version of the restriction clause to ask whether **oxygen** is a natural kind. This shows how this formulation can be used to clarify a particular case and also underscores the connection between the domain of enquiry and the conditions for what counts as success.

In the subsequent section, I consider cases where the domain of enquiry is being shaped at the same time as the enquiry into the domain is being conducted. We can think of kinds in such an enquiry as *fungible* in the sense that a choice of taxonomy might be vindicated by a corresponding change to the objects of enquiry. Because of the restriction clause, fungible kinds are too loose to be natural kinds.

D.1 Example: the domain of chemistry

Examples of oxygen chemistry – both the defeat of phlogiston and the development of atomic chemistry – are tired warhorses of debates about scientific realism. Much of that debate concerns whether Lavoisier's neologism 'oxygen' attached to the same category which our word 'oxygen' now does. This issue of reference need not concern us, since we are interested in the taxonomic category of oxygen as part of the periodic table of elements rather than in the referential continuity of names. I will not attempt to detail the history of the case either.

Robin Hendry argues that the chemical taxonomy in terms of atomic number or nuclear charge best accounts for chemical phenomena, where that domain is designated by the practice of chemistry going back at least to the time of the chemical revolution. He asks how the history might have gone 'if Lavoisier and other scientists had had different interests, or a different background conception of composition' (2010, p. 151). This roughly parallels the question of what would have happened if their enquiry had been directed to a different domain. Hendry answers:

> In fact it is not obvious that anything like modern chemistry would have emerged, because the standards of similarity and difference that shape a discipline influence what it discovers. None of this undermines the fact that nuclear charge is a real physical property that predated chemists' knowledge of it, or the fact that the patterns of behaviour it determines are a genuine feature of the causal structure of the world. There may be other patterns of behaviour in nature, and these may be of interest to other communities of scientists (actual or merely possible), but that is irrelevant to the ... element names as they were actually introduced. (2010, p. 151)

Hendry insists that nuclear charge is a real physical property and that chemists would not have succeeded at the task they set for themselves had they not adopted the taxonomy of elements; this amounts to arguing, as I would put it, that elements are natural kinds for chemistry. This will not satisfy an anti-realist who does not believe in oxygen. What it shows instead is that, insofar as we believe in the stuff *oxygen* or the things *oxygen atoms*, we should believe in the naturalness of the kind **oxygen** for chemistry.

The important qualification Hendry adds is that chemists might have succeeded without that taxonomy, but they would have succeeded *at something else*. If we take the domain of chemistry to be historically circumscribed, then the periodic table satisfies the restriction clause. If we were to drop the requirement that success be *in the specified domain*, then it would not satisfy the restriction clause; the restriction clause would rule out anything Lavoisier discovered as a possible natural kind, because other taxonomies could have allowed chemists to attain success – albeit success at something else.

For this to work, we need to specify what the domain of chemistry is. It may be tempting to give a glib characterization of the domain of chemistry. Alexander Bird (2010, p. 127), for example, quotes Linus Pauling: 'The different kinds of matter are called *substances*. Chemistry is the science of substances – their structure, their properties, and the reactions that change them into other substances' (Pauling 1970, p. 1). Bird is interested in the reference of substance terms like 'oxygen', and he uses this characterization of the domain of chemistry to argue that 'it will be chemical facts that determine the identity of substances' (Bird 2010, p. 127); since chemistry is the science of substances, chemists are the arbiters of substance talk. Yet matters are more complex than that. As Pauling notes just after the passage cited by Bird, the definition of chemistry as the 'science of substances' is inadequate. It is too narrow, because chemists are also interested in energy like visible light (because a substance's having a color amounts to its interaction with visible light) and X-rays (used in diffraction to determine the structure of crystals). It is also too broad, because almost all sciences are concerned with some substances. For example, a 'nuclear physicist studies the substances that constitute the nuclei of atoms', and a 'geologist is interested in the substances, called minerals, that make up the earth' (Pauling 1970, p. 2). Pauling concludes, 'It is hard to draw a line between chemistry and other sciences' (1970, p. 2).

I think we can identify the domain of chemical enquiry in a way that accommodates Pauling's concerns. Electromagnetic radiation – **visible light** and **X-ray** – are not chemistry's central concern. As something at the edge of chemistry's domain, the categories are taken over from optics and the physics of electromagnetism. Similarly, the chemical characterization of minerals is at the edge of geology's concern and so is taken over from chemistry. Some of the central kinds for geology, such as **stratum** and **epoch**, make no appearance in chemistry.

In considering the development of modern chemistry as a discipline, we can also make its domain more precise. Chemists were concerned with specific substances (such as molecules) and phenomena (such as molecular binding). Paul Churchland argues that these substances and phenomena only exist contingently and that their contingency undercuts the claim that chemical kinds count as natural kinds. Even given the actual laws of nature, a hostile universe might preclude there being any molecules. Churchland writes:

> Whether and which elements get formed is highly sensitive to the details of the environment. ... Were the universe-in-general under the same gravitational squeeze that grips the matter of a neutron star, then none of the familiar chemical elements would exist. Our world would be a symphony of purely nuclear, rather than of electron, 'chemistry'. (1985a, p. 13)

Churchland argues further that most of the kinds that are identified by science are similarly contingent and that in a slightly different world there would be different kinds of things. He concludes, 'Either all of these are natural kinds then, or none of them are. I conclude, tentatively, that none of them are. Save only the elect few from the most basic physics, *all* kinds are merely "practical" kinds' (1985a, p. 13). (Although it reflects common usage, there is something wrong about using 'practical' as an antonym for 'natural' in this way; I return to this point in Chapter 4 § A.)

Churchland is right to say that many domains exist only contingently. Would **molecule** be a natural kind for chemistry in Churchland's imagined gravitational-squeeze world? There would no creatures like us in such a universe, of course, nor would there be the phenomena that comprise the primary domain of our chemistry. So we might answer *no*, because there would be no molecules in the world. Alternately, we might answer *yes*, because *chemistry* just is the study of (among other things) molecular phenomena; in the imagined world there just would happen to be no occurrent objects falling within that domain. Neither of these answers is particularly interesting, however. Once we acknowledge that natural kinds are features of the actual world that we hope to identify in our science, facts about them in exotic counterfactual worlds are just a philosophers' game. If you insist on a definite answer, pick whichever of *no* or *yes* you find more congenial.

A similar point can be made for some oddly specified domains. For example, suppose you were to specify the domain of your enquiry to be *the role of phlogiston in respiration*. Would **phlogiston** be a natural kind for that domain? We might answer *no*, because there is no such thing as phlogiston. Alternately, we might answer *yes*, because the domain of enquiry as conceived is concerned with phlogiston. Even if we did let the category **phlogiston** count as a natural kind for this enquiry, however, there still would not be any phlogiston stuff. It would not be a kind of thing in the world, but instead an empty cell in the taxonomy. Empty cells in a taxonomy are at least sometimes legitimate – open spaces in the periodic table, for example. Perhaps we should exclude the proposed enquiry of *phlogistology* as not legitimately scientific, since we know that there is not any phlogiston. Even if we admit it as a possible enquiry, the question is only whether **phlogiston** would be a natural kind for this perversely overspecified enquiry. In any case, **phlogiston** will not be a natural kind for chemistry. (I revisit the issue of whether some enquires might be ruled out as unscientific in Chapter 5.)

Of course, one might object that there was a time when phlogiston seemed to be a natural kind. It figured in the best chemistry before the work of Lavoisier, and it sustained some inductive and explanatory success. Again, this is well-traveled terrain from debates about realism. I am not going to rehearse the historical evidence, but I suggest that phlogiston only ever *seemed to be* a natural kind. The predictive and explanatory failings that led chemists to abandon phlogiston as a theoretical posit also show that it fails to satisfy the success clause. Chemistry was concerned with phenomena such as respiration and combustion, which actually did involve oxygen. Oxygen existed all along and phlogiston never did. So we can say similarly that **oxygen** was the natural kind all along, even before it was identified and named.

There is certainly more to be said, but I hope to have said enough to show how my conception of natural kinds can be applied to the case of chemical elements. Modern chemistry would not have given a successful account of the domain into which it enquired had it not recognized oxygen and the other elements.

D.2 Fungible kinds

The juxtaposition of the natural and the practical points toward a type of boundary case. If the objects which comprise the domain of enquiry are being shaped by the very same project that attempts to describe them, then it is still a live option to construct the domain so that a different taxonomy would fit it. Thus, any taxonomy that could be employed would fail to meet the restriction clause. Querists will successfully employ some kinds, but they will not be natural kinds. Call these *fungible kinds*, because a taxonomy that could underwrite success might be exchanged for a different taxonomy – allowing for comparable success – by making corresponding changes in the domain of objects itself.

To take a humdrum example, suppose my wife and I order a slice of cheesecake and agree that each of us gets half of it. We might just begin eating from opposite ends and continue until it is consumed, but we know from experience that one of us will eat at a different rate than the other, making for an unfair division. So one of us uses the edge of a fork to trace a line across the surface of the cake and says, gesturing to one side of the line, 'This is my half.' The two parts of the cake form separate kinds, **her half** and **my half**. These kinds are practical, because the proof of them is in the eating. Moreover, they are not natural kinds. There was no determinate division before the line was drawn in the cheese. The division might have been made at any angle just so long as it partitioned the cake into two roughly equal zones of control. So there are indefinitely many taxonomies distinguishing her half and my half which would have done equally well, provided there was a corresponding difference in where the actual line was traced. It is for this reason, and not just because the zones are of practical importance, that the zones fail to constitute natural kinds. In the characterization of natural kinds I have offered, the taxonomy of the cake that divides the cake into **her half** and **my half** does not satisfy the restriction clause.

To underscore the contrast, imagine that there is distant island where there is a tradition of couples splitting pieces of cake in a regular way. One of the two traces a line in the cake, but along a specified channel. It would be a scandal for them if the line were to be drawn somewhere else – even if it divided the cake into roughly equal pieces. This is sufficiently ritualized, let us suppose, that each member of the couple could eat just their half even without the line, because every piece is split in the same way. An anthropologist studying these islanders would start to think of their pieces of cake as consisting of *her half* and *his half*. Thinking in this way would allow the anthropologist to make sense of the islanders' behavior in a way that would not be possible without thinking of the cake as bifurcated. So **her half** and **his half** satisfy the success and restriction clauses for the anthropologist's enquiry; that is, they are natural kinds.

If one had a taste for paradox, one might use these two cases to say that **her half** both is and is not a natural kind – but that would be a mistake. Remember that natural kinds are relative to specific domains of enquiry. An anthropologist observing my wife and I could make no explanatory headway by taking too seriously the division between her half and my half, because we might have drawn the line differently and so made for different pieces in *our* cake eating; they are fungible kinds. Contrariwise, the anthropologist observing the island couple would do well to attend to the division between her half and his half, because division is important in *their* cake eating. Different objects, different domains, different natural kinds.

The distinction between natural kinds and fungible kinds can be made even among the same objects by situating them in two different domains of enquiry. A dog breed such as **border collie** may be biologically distinct enough to count as a variety or infraspecies; that is, it might count as a natural kind for some biological enquiry. Whether or not this is so will depend on the details. However, the same kind considered as part of the dog breeder's taxonomy is more problematic. In the 1990s, there were disputes among breeders about whether there should or should not be an American Kennel Club (AKC) breed standard (McCaig 2007). A standard specifies a conformation for border collies, a norm of how they ought to look. Such a standard can be used to assess dogs in shows, but it also guides breeders in selective breeding. Groups like the United States Border Collie Club (USBCC) opposed the adoption of an AKC breed standard on the grounds that border collies are work dogs rather than show dogs; that is, they advocate breeding border collies for temperament, eyesight, physical fitness, and herding behavior rather than for appearance. In the domain of dog breeding, where decisions about the relevant norms are in play at the same time as inductive and explanatory considerations, the category **border collie** is not a natural kind. The objects which fall under the category are being shaped at the same time as the taxonomy which describes the objects. Members of the AKC and USBCC might each achieve inductive and explanatory success using their distinct categories, because the decision to adopt a category is coupled to decisions about how the dogs themselves will be bred. So **border collie** fails to satisfy the restriction clause for dog breeding, and it is not a natural kind for that domain of enquiry. The label *fungible* is apt because AKC and USBCC conceptions might be exchanged given corresponding changes in how dogs are bred.

Fungibility of a kind will sometimes be due to what Ian Hacking has called *looping effects* (1986, 1995, 1999). A looping effect is when the identification of a category leads to transformation in the things that fall under that category. That is just what is happening with border collies.

Hacking is primarily interested in *human kinds*, kinds of people. To take a simple example, consider the fans of a television show. Anthropologists tracking viewing behavior in the 1960s and 1970s might have posited the category **Star Trek fan** to describe a specific demographic. The rubric of *Trekkie* is more than just a demographic category, however. It became a new way for people to be. People who conceived of themselves as Trekkies acted differently and organized together in ways which transformed what it meant to be a *Star Trek* fan. The return of *Star Trek* in the 1980s, as a motion picture and then as *Star Trek: The Next Generation*, was a response to the show's committed fans. So the self-conception of *Star Trek* fans changed both what it meant to be a *Star Trek* fan and what *Star Trek* itself was.

Hacking himself gives the example of **homosexual**. Homosexual behavior can, in some sense, be described among animals and in ancient civilizations. This sense misses much of what we now mean by 'homosexual' or 'gay'. According to Hacking, the term and category *homosexual* was initially introduced 'in medico-forensic discourse late in the nineteenth century' and it was applied 'by the knowers to the known'. He adds, 'However, it was quickly taken up by the known, and gay liberation was the natural upshot. One of the first features of gay liberation was gay pride and coming out of the closet. It became a moral imperative for people of the kind to identify themselves, to ascribe a chosen kind-term to themselves' (1995, p. 381). A result of the concept's being available is that *homosexual* became something that people could conceive of themselves as being. It became something they could publicly declare and shape their lives around. Gay pride parades were about crafting the gay identity rather than merely about celebrating the bare biological act of gay sex. What gay men are now is thus conditioned by theorizing about earlier gay men. (Perhaps **homosexual** would not be a natural kind in any case. That cannot be established without more attention to the details. Similarly, whether the looping effect undercuts its status as a natural kind depends on details such as whether alternative categories with their corresponding looping effects could have produced paths to success.)

Thomas Reydon (2010) argues that artifacts are more likely to be subject to looping effects than categories of people. In general, for objects which humans can causally affect, the kinds which include those objects may be subject to looping effects – and kinds which are subject to looping effects may thereby fail to be natural kinds. They become fungible rather than natural kinds.

However, not every kind within the human sphere is actually subject to looping effects. Moreover, the difference between a natural kind and a fungible kind depends on the relevant domain of enquiry. To return to an earlier example, the kind **border collie** might be a natural kind for biological enquiry as separate from the actual management of dog breeding. The case is even clearer if we consider a twenty-first-century historical description of nineteenth-century dog varieties. How we categorize the dogs that lived in the past might affect the dogs which will live in the future, but our description cannot make a causal difference to the organisms that lived and died more than a century ago. Perhaps nineteenth-century dog varieties form natural kinds, but perhaps they do not; the matter will depend on the details, but not on looping effects.

(One might worry that matters wrought by human artifice cannot correspond to *natural* divisions. Humans are part of nature, though, so we should not insist that artificial necessarily contrasts with natural. I argued this at greater length in Chapter 1 § B.5.)

Scientific instruments as fungible kinds

Although standards vary between disciplines, some scientific papers provide considerable detail about the instruments that were used in the work. Even brand names and model numbers may be mentioned. In a paper about contemporary replication of Jean Perrin's work on Brownian motion, for example, the authors write, 'Our experiment was performed using an American Optical Microstar trinocular microscope. The 100× objective was chosen to maximize displacement of the microsphere image at the camera's CCD array. ... The computer communicates with the camera via a FireWire interface'

(Newburgh, Peidle, and Rueckner 2006, p. 479). A recent paper in marine biology gives similar detail:

> In situ observations and video recordings of spawning *O. rubiplumus* females on whale-1820 were made with a high-definition digital video camera (Panasonic WVE550) aboard the ROV *Tiburon*. Laboratory observations of and digital video recordings (with a Canon Powershot G9) of streams of zygotes being spawned by females of Osedax 'yellow collar' were made at SIO. (Rouse et al. 2009, pp. 396–7)

The recently discovered Osedax bone worms are biologically interesting and important, and so it makes sense to see the paper which contains this description as part of successful science. If we naïvely suppose that all of the categories that appear in successful science of a domain should count as natural kinds for that domain – i.e. if we only require the success clause – then it would follow that **Canon Powershot G9** is a natural kind for marine biology.

There is no *in principle* objection to something like digital cameras or even a particular brand of digital camera forming a natural kind. It may be that **camera**, **digital camera**, or even **Canon Powershot** would be natural kinds for economics, history of technology, or some other enquiry. The absurdity, however, is to think that they should be natural kinds *for biology*.

The important thing to note is that the specific instruments are inessential to the results. They are mentioned because the precise results were produced in that way. The scientifically significant aspect of the work does not depend on the exact choice of instruments, however. The video recorded by the marine biologists can be viewed on their website, and a different camera would have made some difference to the exact stream of bits that was produced and is posted there. Yet there is no magic power peculiar to the Canon Powershot that made the observations possible. The group might have used a different camera and achieved comparable biological success. Since they had a choice of cameras and success could have been achieved for any of several choices they might have made, **Canon Powershot** would count as a fungible kind in their enquiry.

If a definition of natural kinds had only the success clause, then it would not distinguish **Osedax bone worm** as a natural kind for marine biology from **Canon Powershot G9** which is not. To put the point differently, it could not distinguish natural kinds from fungible kinds. The restriction clause allows us to do this, allowing us to say that – despite being referred to in successful science – the specific kinds and brands of instruments are not natural kinds for the enquiries in which they are employed.

What happens next

There are two ways that the discussion will go from here. One direction is to apply the conception of natural kinds to specific cases; that is what I do in Chapter 3. Another direction is to situate the conception with respect to various philosophical debates; I turn to that in Chapter 4. The two threads merge again in Chapter 5 (where I consider some examples and objections that arise from them) and Chapter 6 (where I consider examples of homeostatic property cluster kinds).

3
Natural Kinds Put to Work

In the previous chapter, I argued that a natural kind in a domain is (a) part of a taxonomy that allows enquiry in the domain to achieve inductive and explanatory success, (b) such that success would not be attained with a taxonomy that excludes the kind. I call these criteria the *success clause* and *restriction clause*, respectively.

In this chapter, I apply the conception of natural kinds to specific cases from astronomy (whether Pluto is a planet, § A), biology (whether species taxa and the species category are real, § B), and cognitive science (whether distributed cognition is cognition, § C). In cases like these, we are considering a community in which science is already underway. So we can restate the conditions in this way: The community has identified natural kinds for a domain insofar as (a) the community's enquiry in the domain achieves inductive and explanatory success using a taxonomy that includes those kinds, and (b) the community would *not* have attained success had it *not* employed a taxonomy that includes those kinds.

A. Eight planets, great planets

At an August 2006 meeting in Prague, the International Astronomical Union (IAU) adopted a definition of 'planet' which excluded Pluto (IAU 2006). There was a good deal of hubbub about it at the time, with blog posts and news items decrying the end to the canonical list of nine planets that we all learned in grade school. In part because of this outcry, it was also hailed as a triumph of science over sentimentality. 'Capping years of intense debate,' wrote one reporter, 'astronomers resolved today to demote Pluto in a wholesale redefinition of planethood that is being billed as a victory of scientific reasoning over historic and cultural influences' (Britt 2006).

Did this reflect the natural kind structure of astronomy? Or was it just a decision by astronomers as to how they would use words?

Many conceptions of natural kinds would have us say the latter. Planets are subject to gravitation and not some special planet-action law, so identifying natural kinds as the things that appear in basic laws of nature would rule out **planet**. They are made of the usual chemical elements and not of special planet-stuff, so distinguishing natural kinds by their intrinsic properties or smallest constituents would also exclude **planet**. I argued against those approaches in Chapter 1, but the IAU decision provides a further reason to reject them: What we want to know is whether and to what degree the IAU's taxonomic decision reflected the structure of nature. By ruling out **planet** as a natural kind for entirely general reasons, such approaches would make *natural kinds* irrelevant to the question of how natural astronomy's kinds are.

I also argued in Chapter 1 that we should not treat the theory of reference as a constraint on our account of natural kinds. Still, one might think that the Kripke–Putnam account of reference could give us leverage on the fate of Pluto. If the word 'planet' were a rigid designator, then there would be a determinate fact of the matter as to whether Pluto was a planet or not. One could say that the IAU decision was correct if it got the metaphysics right. Recall that, on the Kripke–Putnam story, the reference of the word planet depends on how it was introduced. We imagine a primordial language user pointing to Mars and saying, 'Lo, it is a planet.' Thus dubbed, 'planet' would mean *that kind of thing*. This would hold in all possible worlds, so it would have held in Prague, August 2006.

This story might be appealing for premodern astronomy. The word 'planet' descends, etymologically, from the Greek word for *wanderer*. When the Earth was thought to be at the center of everything, the planets were the points of light that moved against the background of the fixed stars. This descriptive content would peg planets as wandering stars, but the triumph of heliocentric astronomy involved recognizing that Earth itself was also a planet. Thinking of 'planet' in the Kripke–Putnam way, we can treat this as a discovery rather than as a change in the meaning of the word. Earth is the same kind of thing as Mars, in the relevant respect, and that was so in ancient times regardless of Ptolemaic doctrine to the contrary.

The Kripke–Putnam idea that the word is connected to the kind by a metaphysical leash just restates the problem about Pluto, however, because what we want to know is whether Pluto is *that kind of thing*. Kripke and Putnam treat kinds like **water** and **tiger** as being distinguished by internal structure, but there is no material composition common to the planets. So we need to think through the systematic constraints of astronomy before we can answer the question of whether 'planet' referred to Pluto or not. That is just to say that we need to solve the problem of whether **planet** is a natural kind – and, if so, which natural kind it is – before we can trace the tether that connects it to the word 'planet'. Just as I suggested in Chapter 1, the issue of reference is separate from and subsequent to the question of natural kinds.

I begin with some historical context, discussing developments in the eighteenth and nineteenth centuries that set the context for the discovery of Pluto (§ A.1). Then I discuss the discovery of Pluto itself and conditions that ultimately forced the IAU to make a decision (§ A.2). With the background in place, I turn to the physical and astronomical considerations that underwrite the IAU decision (§ A.3). Finally, I consider several objections to the IAU definition. Although the definition is not as meticulous as it might have been, I argue that it picks out a natural kind (§ A.4).

A.1 Numerology and asteroids

Our story can begin with the Titius–Bode rule. First proposed in 1766 to accommodate the six known planets, the rule is a procedure for generating a series of numbers. The values in the series roughly fit the size of the planets' orbits. When William Herschel discovered Uranus in 1781, its orbit roughly fit the eighth value in the series. Although Herschel had not set out with the rule in mind, Johann Bode was happy to see the discovery as confirmation of the rule.

In the numerical series, Mars corresponds to the fourth number in the series and Jupiter corresponds to the sixth. This suggested that there was a missing planet, and that the size of its orbit should approximate the fifth number in the series. A systematic, collaborative effort to survey the sky was mounted and discovered *something* in 1801, roughly at the right distance from the Sun to be Bode's missing planet. The object, dubbed Ceres, is smaller than the other planets. Moreover, other comparable objects were found in its neighborhood: Pallas in 1802, Juno in 1804, and Vesta in 1807. Some people spoke of these as planets, but Herschel suggested that they should be considered a different kind of thing. He suggested the name *asteroid*. (Etymologically, 'asteroid' means *starlike*. We should not take this etymology too seriously, however. Nobody ever thought that Ceres was more like a star than it was like a planet.)

Meanwhile, careful observation of Uranus revealed discrepancies between its orbit and predictions of its orbit. The difference could be explained by the presence of an unknown massive body in the outer Solar System. Urban Le Verrier made a prediction of where such an object would be. In 1846, an object was discovered within one degree of the location in the sky where Le Verrier had predicted the planet to be. The object, dubbed Neptune, is much larger than the asteroids. It was a discovery of sufficient gravity – in both senses – to count as a planet.

Any uncertainty about the inferior status of the asteroids was settled as more asteroids were discovered: a fifth in 1845, three more in 1847, and hundreds by the end of the century. It made sense to treat the asteroid belt as a group of objects, none of them on par with the eight planets (now including Neptune).

A.2 Enter Pluto

Adding Neptune to the calculations did not resolve the entire discrepancy between observed and calculated positions for Uranus. Some astronomers, trying to mirror the triumph of Le Verrier, posited further planets to account for the difference. One of these astronomers was Percival Lowell, a former businessman and diplomat. He founded Lowell Observatory with his own money, and one of the tasks at the observatory was to find the missing Planet X. After Lowell's death in 1916, workers at the observatory continued the search. In 1930, Clyde Tombaugh at the Lowell Observatory spotted *something*. It was roughly at the distance Lowell had predicted, although Tombaugh's search had covered a considerable swath of the sky. The object, dubbed Pluto, came to be accepted as a ninth planet.

Yet Pluto was only approximately where Planet X had been predicted to be, and it is much too small to make much gravitational difference. It does not suffice to explain any alleged discrepancies in the orbit of Uranus. In subsequent decades, the case for Pluto became more dire. Further observations suggested that it was even smaller than previously believed. In 1978, James Christy and Anthony Hewitt discovered Charon, a companion to Pluto. Thinking of Pluto as a planet, Charon is its moon. Yet Charon is about one half the diameter and one tenth the mass of Pluto. The center of gravity of the Pluto–Charon system is between them, so it would not be too pedantic to say that *the two orbit each other* rather than saying that Charon orbits Pluto. (For comparison, Earth's moon is about one quarter the diameter and one percent the mass of Earth. The center of mass of the Earth–Moon system is inside the surface of the Earth.)

In 1980, this ignominious trend for Pluto led two astronomers to plot the estimates of the mass for the ninth planet over time (see Figure 3.1). They fitted a curve to the observations and predicted that, if the trend continued, Pluto would reach zero mass in 1984. Beyond that point, the curve predicts a complex mass with a negative real component. 'One can push mathematical extrapolations too far,' they quip. '[P]erhaps there is a physical explanation for this disappearing act' (Dessler and Russell 1980).

The banal explanation is that the position data of Uranus had not been sufficiently precise to sustain the kind of prediction Lowell tried to make. Astronomers who predicted a ninth planet on that basis were, as David Weintraub comments, 'interpreting those data with a subtle bias they did not recognize they had' (2007, p. 146). This common bias, along with the prestige of the Lowell Observatory, led to the acceptance of Pluto as a *planet* even though it never quite fit in with the others.

There was no pressure to expel it as long as it was a singular oddity. Nine planets? Fine.

Just as the asteroid Ceres was treated as a different kind of thing – not a planet – once others of its ilk were discovered, Pluto's uncomfortable membership in the league of planets became unstable when astronomers found more trans-Neptunian objects like it. This 'vast swarm of objects beyond Neptune' (Brown 2006) is often called the Kuiper Belt, after the astronomer Gerard Kuiper. It is like a trans-Neptunian counterpart to the asteroid belt. Just as the company of other asteroids cost Ceres its planethood, the company of other Kuiper Belt plutoids would force the question with Pluto. The first Kuiper Belt object (KBO) to be observed, after Pluto, was discovered

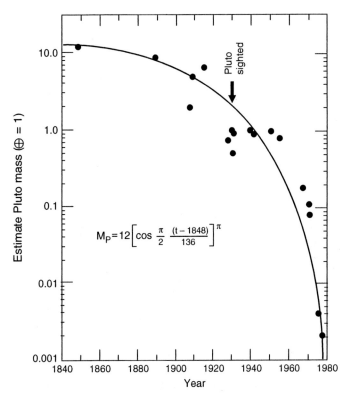

Figure 3.1 Estimates over time for the mass of a trans-Neptunian planet and later for the mass of Pluto so-called. A curve fitted to the estimates predicts zero mass in 1984. (Adapted from Dessler and Russell (1980). © 1980 American Geophysical Union. Reproduced/modified by permission.)

by astronomers David Jewitt and Jane Luu in 1992. Numerous other KBOs were discovered in the 1990s, and in 2003 a survey at Palomar Observatory discovered a KBO that is 27 percent more massive than Pluto. The object, later named Eris, brought the issue to a head.

A.3 The constraints of astronomy

The IAU resolution which passed in Prague set three conditions for planethood: 'A planet is a celestial body that (a) is in orbit around the Sun, (b) has sufficient mass for its self-gravity to overcome rigid body

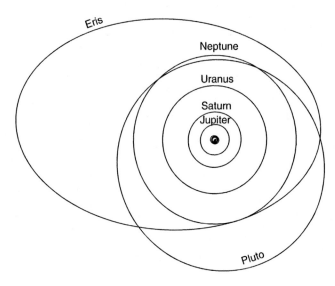

Figure 3.2 A sketch of our solar system, showing the orbits of the outer planets, Pluto, and Eris. (Based on the diagram 'Tilted Eris' (NASA 2011).)

forces so that it assumes a hydrostatic equilibrium (nearly round) shape, and (c) has cleared the neighbourhood around its orbit' (IAU 2006). The resolution just defines what it is to be a planet in our Solar System. A general definition which covers planets around any star would have to add a further condition so as to exclude brown dwarfs and companion stars in binary star systems. This complication is not relevant to the case of Pluto, so I will set it aside and just consider the three criteria in the IAU definition.

a. *A planet's primary orbit must be around a star*

This point is straightforward. The Jovian moon Ganymede is larger than Mercury, but it counts as a moon rather than a planet.

b. *A planet must be big enough to be round*

Asteroids tend to be lumpy and irregularly shaped. A large enough object tends to be round, however, because its own gravity causes it to collapse and pull it together into a ball. So condition (b) is motivated by considerations of physics and marks a significant difference between planets and other things. Mike Brown (one of the astronomers who discovered Eris, but not one of the voters in the IAU decision) explains, 'This transition from irregularly shaped to round objects is important in the solar system, and, in some ways, marks the transition from an object without and with interesting geological and planetary processes occurring' (2006). Given my conception of natural kinds, it is plausible to say that this criterion corresponds to the boundary of a natural kind.

Pluto is large enough to be round, but so are Eris and some other recently discovered Kuiper Belt objects – and so is the asteroid Ceres.

c. *A planet must clear its neighborhood*

A planet is the significant thing in its orbit, whereas Ceres is just one of many asteroids and Pluto is just one of many KBOs. 'As planets form,' explains science writer Fraser Cain, 'they become the dominant gravitational body in their orbit in the Solar System. As they interact with other, smaller objects, they either consume them, or sling them away with their gravity' (2008). This is a straightforward idea, but there is neither a threshold for how clear the neighborhood must be nor a boundary line around what counts as a planet's neighborhood. Brown admits that 'the actual wording of the definition is not as precise as it might have been, giving people room to quibble and to say that the definition is unclear' (Brown 2006).

David Weintraub worries that a criterion like this fails to distinguish obvious planets from non-planets. Jupiter's orbit is crowded with 'clouds of objects called Trojan asteroids', Weintraub explains. 'These objects span a range in longitude that covers about a forty-five degree arc centered on each of the L4 and L5 [Lagrange] points; that is, the Jovian Trojan asteroids fill about 25 percent of Jupiter's orbit. Jupiter, therefore, is the largest object in the Jovian ring' (2007, p. 205). Ceres is part of the asteroid belt and Pluto part of the Kuiper Belt, but Weintraub insists that there is no reason to distinguish those 'wide, filled rings' from the 'narrow, unfilled ring' of Jupiter and its Trojan asteroids. Appeal to such a difference, he maintains, would be 'a flawed means for determining whether an object is a planet, as it does not make reference to the physics of the object itself' (2007, p. 206). So, Weintraub argues that criterion (c) is either astronomically capricious (because it is unphysical) or extensionally inadequate (because it would exclude Jupiter). In order to evaluate this, we need to know a bit more about Trojan asteroids.

There are five points along or near a planet's orbit where the gravitational influence of the planet is balanced by the gravitational influence of the Sun, so that a third, less massive object could hold its position there. The existence of these points was first calculated in the eighteenth century by Joseph-Louis Lagrange, and they are called *Lagrange Points* (see Figure 3.3). They can be used as parking spots for spacecraft; for example, NASA has a solar observatory at Earth's L1 point. There are two stable Lagrange Points, leading and trailing the planet on its orbit by 60°.

Trojan asteroids orbit near the L4 and L5 points, the two stable Lagrange Points. Jovian Trojans were first observed in 1906, and there are now more than 4,000 known (IAU 2010). Martian Trojans were first observed in 1990, Neptunian Trojans in 2001, and now several of each have been discovered. Detecting them is difficult, so there are probably many more that have not been observed. The quantity of Trojans may make it seem as if these planets have failed to clean their neighborhoods. One imagines them to be like the

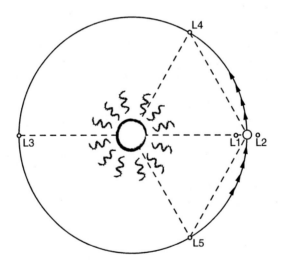

Figure 3.3 Lagrange Points. A star–planet system has five Lagrange Points at which a much lighter object could be parked. Only two of these (L4 and L5) are stable.

Peanuts character Pigpen, with clouds of dirt leading in front and trailing behind. This impression would be mistaken.

When he discovered the eponymous points, Lagrange was investigating the three-body problem in mechanics. The general problem is intractable, so he was considering the special case in which one of the three bodies has a mass that is negligible in comparison to the other two. The calculation ignores both the gravitational effect of the third body and the effect of any other masses. This mathematical idealization is what defines the Lagrange Points. In order to identify the Lagrange Points, then, we must think of a system as comprising two bodies. The Lagrange Points are then places where a third body might be placed. The Earth–Moon system has Lagrange Points, because the two are the gravitationally dominant bodies in the neighborhood. The Sun–Earth system also has Lagrange Points, far enough away that Earth's moon can (to a first approximation) be disregarded.

Note that Trojan asteroids are ones that orbit near the L4 and L5 Lagrange Points. To be more precise, Jupiter's Trojan asteroids orbit near the L4 and L5 Lagrange Points of the Sun–Jupiter system. The very applicability of the label *Trojan* requires that we consider Jupiter to be the dominant mass in its neighborhood, because it depends on the Lagrange Points which in turn result from the physics of Jupiter in relation to the Sun.

An asteroid's being part of the Kuiper Belt is simply a matter of its being in the right part of space. Specifying that region of space does not require any reference to Pluto or Eris. This makes the situation of Jupiter and its Trojan chaff very different than Pluto or Eris and the other bodies that contingently share the Kuiper Belt with them.

This suffices to answer Weintraub's objection. Moreover, it illustrates why the criterion plausibly marks the boundary of a natural kind. Recognizing Jupiter as a planet is crucial to understanding why the Jovian Trojans are arranged as they are – they are clustered around the planet's L4 and L5 points. If astronomers attempted to proceed without the taxonomy of planets and Lagrange Points, then the 'narrow, unfilled ring' in Jupiter's orbit would be inexplicable.

Pluto crosses the orbit of Neptune, and this might also seem to be a *prima facie* counterexample to Neptune clearing its orbit. Just like the Trojans, however, this special case is also explicable in relation to Neptune's gravitational influence. Neptune's moon Triton is, roughly speaking, another Pluto or Eris that got close enough to Neptune that it was captured. If Pluto crossed Neptune's orbit at some random point, it would eventually either be slung away by Neptune's influence, be captured as another moon, or collide with Neptune. Yet Pluto does not cross Neptune's orbit haphazardly. Rather, Pluto orbits the sun twice for every three times that Neptune orbits the sun, and this resonance means that they do not collide. Moreover, Neptune's influence acts to keep Pluto in the resonant orbit. If Pluto speeds up or slows down, then Neptune's gravity pulls it to a higher or lower orbit so as to maintain the 3 to 2 ratio. We understand the way that Pluto passes through Neptune's neighborhood only when we consider Neptune to be the dominant influence there.

A.4 Natural kinds and the fate of Pluto

In the previous section, I argued that all the conditions in the IAU definition of 'planet' pick out features that are astronomically significant. Identifying the category thus furthers the systematic success of astronomy. So, **planet** satisfies the success clause for being a natural kind. Moreover, I argued that adopting a weaker definition of 'planet' would have come at a cost in explanatory success. So there is reason to think that **planet** satisfies the restriction clause as well.

In this section, I consider three objections to counting **planet** as a natural kind. They will reveal the extent to which the IAU decision was not entirely determined by the world, but also the sense in which it was a response to what the astronomical domain is like. The latter sense is sufficiently strong that we can say, according to my way of thinking about natural kinds at least, that the definition identifies a natural kind.

Objection 1:   The IAU definition is a mismatched laundry list of criteria

One might object that the IAU definition is hodgepodge of three different criteria put together to solve a political problem. I argued earlier that the contentious criteria do in fact correspond to astronomically important factors, but why should the *conjunction* of the three criteria distinguish anything important? Note also that the definition the IAU actually adopted just settles what we will call a planet in *our* solar system; a real astronomical kind should hold for planets around any star. The resolution was (the objection concludes) simply a veneer of objectivity for a decision about what to do about the beloved and reviled Pluto.

There are several replies to this objection:

First, the domain of enquiry we are considering is not astronomy *tout court* but instead the astronomy of solar systems. For this domain of enquiry, criterion (a) is not arbitrary; it is almost trivial. One might protest that we only enquire into this domain at all because we live in a solar system. Scientists who grew up in the vast interstellar expanses might not care about such things. Nevertheless, the domain (once specified) determines its natural kinds. The fact that this domain is one we care about does not make these kinds any less real. I discuss this point at greater length in the next chapter.

Second, consideration of how a third body will interact with the sun and a planet is a special case of the three-body problem. The fact that the three-body problem is insoluble in general is an important mathematical constraint and so, for computationally bounded agents, special case solutions provide an important affordance. The IAU definition of planet exploits this. If the sun were surrounded by a uniform ring of small particles, then the approach would not apply. Yet it does apply, and it does so because of features of the solar system.

Third, criteria (b) and (c) are not independent. Both are a matter of the object in question being *sufficiently massive*. It is plausible, in fact, that anything which satisfies (c) will necessarily satisfy (b); i.e. anything which was not massive enough to be round could not possibly clear its orbit. The criterion (b), roundness, is mentioned separately for political reasons; namely, that it was a separate criterion that had been discussed in the debate leading up to the IAU decision.

With that in mind, we might more compactly define a planet in this way: *A planet is an object which is not itself a star but which is massive enough to dominate its orbit around a star.*

This definition is equivalent to the IAU definition, but it makes the explanatory power of thinking about *planets* more perspicuous. Knowing that something is so massive allows us to explain why it is round and to predict what kinds of things will be in its orbit: Other things will either collide with the planet (and so add to it), be slung away by the planet's gravitational influence, be shepherded to the planet's L4 and L5 Lagrange Points, be pulled into orbit around the planet (as a moon), or be locked into a resonant orbit. This disparate lot of possible fates is explained in a unified way by the planet's dominant mass.

Fourth, there is good reason to think that many or most stars will have planets. When a star forms, there is a lot of other stuff around it. Although the stuff may begin as homogenous rings, particles collide and accrete. Once they accumulate into a minimal planetoid, they begin to affect their neighborhood in a way that furthers the development of full-fledged planets. Admittedly, direct observational evidence for this comes almost entirely from our own solar system. Although astronomers have recently been able to identify exoplanets (planets orbiting other stars), the exigencies of distance make it impossible to determine whether (for example) they have Trojan asteroids; it is difficult to observe Trojans even in our own solar system. Because of these empirical limitations, it would overreach to say too much about what planets must be like in general. Nevertheless, there is a good reason to think that planets are not just an idiosyncratic feature of our star system.

In sum, the orbit-domination of planets allows us to explain facts about the formation, development, and present configuration of the solar system. The objection which alleges the IAU criteria to be a disparate checklist fails. The kind **planet** is a natural kind for astronomy. This is a fallible judgment, of course, but any real example must be.

Noting that the IAU definition is formulated in terms of distinct criteria, though, we can ask about things that satisfy some but not all of them. The same IAU resolution which defines 'planet' also defines the term 'dwarf planet' to mean an object which meets conditions (a) and (b) but not condition (c); i.e., a dwarf planet orbits a star and is large enough to have pulled itself into a round shape, but is not large enough to have cleared its orbit. Pluto, Ceres, and Eris are all dwarf planets. Condition (b) depends on the physics of the object and determines some astronomically interesting features. So it is plausible that **dwarf planet** is a natural kind. It is just a different natural kind than **planet**.

There might be interesting objects which satisfy (b) and (c) but not (a); i.e. objects which clear their orbit but which orbit something besides a star. Perhaps our own moon is such a thing. Earth itself explains the development and structure of things near its orbit around the star, but the moon might explain more local facts about the development and structure of things near its orbit around Earth. Note that many moons are too small to satisfy (b), and astronomers have no special term for round moons. This suggests, although not decisively, that the category has minimal explanatory value.

Objection 2: 'Planet' is dispensable, because narrower kinds might do the same work

One might argue that a more narrowly defined kind would do the same explanatory work that **planet** can do. Despite the common feature of dominating their respective orbits, the planets are a heterogeneous bunch. Mike Brown writes:

> If you were to start to classify things in the solar system from scratch, with no preconceived notions about which things belong in which categories, you would likely come to only one conclusion. The four giant planets – Jupiter, Saturn, Uranus, Neptune – belong in one category, the four terrestrial planets – Mercury, Venus, Earth, Mars – belong in one category, and everything else belongs in one or maybe more categories. (2006)

Attending to composition, the giant planets and the rocky planets are very different. So, one might object, astronomy should have kinds **gas giant planet** and **rocky planet**. Since a gas giant planet is necessarily a planet, whatever predictions follow from something's being a planet must follow from its being a gas giant planet; likewise, for rocky planets. Once these kinds are in place, then, the kind **planet** turns out to be superfluous.

The objection fails, because such economy of kinds would come at a cost in explanatory power. The thing which allows a planet to have Trojans does not depend on its composition. Mars and Neptune both have Trojans, a possibility opened up for the same reasons in each case – namely, that they are the dominant masses in their respective orbits. If we considered them to be members of disparate kinds, Neptune as a gas giant and Mars as a rocky planet, then their important similarity would disappear.

Moreover, there may be objects in deep space that are structurally like gas giants. If a cloud of gas collapses but lacks sufficient mass to initiate fusion, then the result would be something like an orphan Jupiter. Such objects do not lend themselves to direct observation, but – if we found one – it would make sense to consider it to be the same kind of thing as Jupiter. So systematic scope suggests the kinds for astronomy should be **gas giant** and **planet** rather than their intersection. Those kinds together encode what is common between Jupiter and a failed unter-star (that they are gas giants) and also what is common between Jupiter and Mars (that they are planets).

Objection 3: There might still be criteria that preserve the list of nine planets

The IAU's decision about the *word* 'planet' does reflect the history of the term better than a definition which only required (a) and (b) – a definition which would have included the dwarf planets such as Pluto. The decision not to count Ceres as a planet was made and settled in the nineteenth century. Allowing Pluto to remain a 'planet' by adopting a definition only in terms of (a) and (b) would have required upsetting that precedent and reclassifying Ceres. Moreover, as I suggested above, the extension of the IAU definition is a class of things that play an important role in the development and structure of a star system.

Nevertheless, one might object, I have not shown that there *could not be* any astronomical considerations which would distinguish the traditional nine planets from all the other things orbiting the sun. It is not clear what the criterion would be, since neither gravitational influence nor composition will do the trick. Mike Brown writes, 'The previous nine ... "planets" encompassed the group of giant planets and the group of terrestrial planets and then awkwardly ventured out into the Kuiper belt to take in one ... of the largest of those objects. Using the word in this way makes no scientific sense whatsoever' (2006). Given our best understanding of astronomy, there is no natural kind in the domain which includes exactly the traditional nine. He might be wrong, of course, but fallibilism should not be an excuse to discuss natural kinds only at airy levels of abstraction.

Even absent any discernible criterion, one might still insist that there could be some metaphysical difference between the nine objects we had considered planets and everything else. There is at least the difference that we called those nine things 'planets' back in 1990 – but that is just an expired, nominal difference. It might matter to history, but not to astronomy. There might, I concede to the objection, be some further ontological difference written undetectably in the fundament. I have nothing to say about fundamental metaphysics, a silence which I will defend explicitly in the next chapter, and as a philosopher of science I have no interest in a difference that only a god could see.

In explicating a word, prior usage counts for something. Although Pluto was called a 'planet' before, the precedent of Ceres (which counted as the largest asteroid rather than a planet because more objects in the asteroid belt were discovered) suggests that 'planet' should not include Pluto now that it is known to be one among many Kuiper Belt objects. Regardless, the precise fate of the word 'planet' is not the central point of the story for us. As Matthew Slater suggests, 'perhaps all the discovery of Eris (and its siblings) does is offer us a choice; either we have fewer planets than ordinar[il]y thought or far, far more' (unpublished). The word 'planet' might have been used in a different way – for the union of **planet** and **dwarf planet**, for example, with new terms invented to cover the more specific categories. Indeed, as Mike Brown (2010) recounts, there were factions in the IAU which had lobbied for a different outcome. There are several natural kinds at play, and astronomers distinguish planets, dwarf planets, gas giants, and so on.

The crucial point is that no scientifically viable choice would have preserved the familiar list of nine planets. The natural kind that includes Mars, Jupiter, and Pluto also includes objects like Ceres and Eris. The natural kind that includes Mars and Jupiter but does not include Ceres or Eris also excludes Pluto. If 'planet' is to pick out a natural kind, then there could be eight or more than a dozen 'planets' so-called in our solar system – but not nine.

B. The abundance of living things

The *species problem*, a stock issue for philosophy of biology and a theoretical concern for practicing biologists, is a cluster of issues about the status of kinds of living things. Reviewing a book on the species problem, the geneticist John Brookfield laments 'too many words chasing too few ideas' (2002, p. 108). The rivalry between various species concepts is not a scientific dispute at all, Brookfield insists. Rather, it is a matter of 'choosing and consistently applying a convention about how we use a word. So, we should settle on our favourite definition, use it, and get on with the science' (2002, p. 107).

Is Brookfield right, and is the only issue one about how scientists decide to use the word 'species'? Or does species talk reflect the natural kind structure of the biological world?

Brookfield is blunt about what he takes to be '[t]he essence of the "species problem"' (2002, p. 107), but there are at least three distinct issues lurking under the heading. Following Richard Richards (2010, pp. 10–11), we can call these the problems of realism, pluralism, and individualism. For the remainder of this chapter, I will only use 'realism' and the other labels to mean aspects of the species problem; I take up *realism* in a broader sense in the next chapter and *pluralism* in a broader sense in Chapter 5.

First, realism is the problem of whether individual species are natural kinds. The species *Homo sapiens*, we tend to think, is more of a feature of nature than the collection of people who live within the city limits of Albany. Nominalists suggest instead that these collections are just groupings that we find useful or not, and so neither is more real than the other. The kind **resident of Albany**, although it might be useful for urban planning, has no significance for biology – yet *usefulness* is a red herring. A natural kind can be useful without giving up its credentials as natural. (I return to this point in Chapter 4 § B.) The crux of the matter is whether *Homo sapiens* is a natural kind for biology. We can ask the same question for other particular species.

Second, pluralism is the problem that biologists identify and characterize species in different ways. This is not a worry about particular species, like *Homo sapiens*, but about the category **species** itself. Ernst Mayr (1964 [1942], ch. 5) lists five species concepts used by biologists, Richard Mayden (1997) lists over twenty-two, and John Wilkins (2009, p. 198) lists 'twenty-five or so'. The plurality – regardless of the exact number – suggests that there is no unified kind **species**.

Third, individualism is the claim that particular species are cohesive individuals rather than mere sets. The former view is that a species forms a single whole, and organisms are its parts; the latter view is that a species is a class or collection, and organisms are its members. The question of individualism can only arise if we have accepted that particular species are real, but it does not require that species all be part of a unified **species** category; it requires realism but is compatible with pluralism. The view that a species like *Homo sapiens* is an individual is often taken to be incompatible with the view that it is a natural kind, but note that this does not follow for the conception of natural kinds I have been advocating. If recognizing *Homo sapiens* is required for scientific success in a domain, then the species is a natural kind regardless of what it *is* at the level of deeper metaphysics. I will say more about individualism in Chapter 6 § D, but set it aside for now.

Taking up the first two issues, the general questions become: Are particular species taxa natural kinds for biology? Is the general category **species** a natural kind for biology?

The species problem is the subject of many book-length discussions, and I cannot hope to resolve it here. Nevertheless, I hope to show that thinking about natural kinds using the conception developed in the last chapter can shed some light on this much-discussed bugbear.

B.1 Particular species (buzz, buzz)

The only way to show that specific species taxa constitute natural kinds is to look at cases. For any particular species, the only way to show that it is a natural kind is by looking at the predictive and explanatory leverage it provides. So consider mosquitoes.

Biologists have discovered that not all mosquitoes can serve as vectors for malaria. Only species of the genus *Anopheles* – and only some of them – can do so. For example, female mosquitoes of the species *Anopheles gambiae* are the primary vector of malaria in Africa. Recognizing these species as distinct taxa is crucial for understanding and responding to malaria. Explanation and prediction are made possible by that recognition, and so the species are (plausibly) natural kinds.

This shows that mosquito species are natural kinds for a domain of enquiry, like epidemiology and public health, which includes malaria. A domain which includes human health and well-being will have **malaria** as a natural kind, and understanding the spread of malaria across human populations requires recognizing the role played by mosquitoes. So, if the domain did not include mosquitoes at the outset, inductive and explanatory demands would require that they be added.

This suggests several ways one might try to resist the example:

First, one might object that we care about public health and malaria deaths only because of our interest in human well-being. Yet a kind's being useful to us does not preclude its being natural, and a domain's being of concern to us does not preclude the domain's having natural kinds. I deal with this point at some length in the next chapter.

Second, if all mosquitoes could transmit malaria, then perhaps the kind **mosquito** would just be one monolithic natural kind for public health. This contingency, one might object, undercuts the claim that particular mosquito species are the natural kinds. To reply: The objection just amounts to saying that, if the structure of the world had been different, then which kinds are natural kinds might be different too. I do not see how it could be otherwise. Again, see the next chapter.

Third, one may worry that we do not ordinarily think of epidemiology as the discipline that identifies species. What we want to know, in order to address the species problem in its realist form, is whether they are natural kinds for biology. One may worry that entomology, which begins with mosquitoes and not human health in its domain, will have little explanatory place for the fact that some species of *Anopheles* can transmit malaria.

In terms of their gross, observable characteristics, there is little to distinguish *A. gambiae* from similar mosquito species. It is part of 'a species complex of seven recently diverged, morphologically identical sibling taxa, including another major malaria vector' (Neafsey et al. 2010, p. 514). Despite this superficial similarity, *A. gambiae* counts as a distinct species because of genetic, reproductive, ecological, and behavioral differences. Moreover, the species is differentiating further. Recent research shows it to be 'composed of at least two morphologically identical incipient species known as the M and S molecular forms' (Neafsey et al. 2010, p. 514). The M and S forms are distinguishable by differences in their ribosomal DNA. Although their ranges overlap, they rarely interbreed. The larvae of the two forms are adapted to different conditions. The S form larvae develop quickly, allowing them to do better in arid areas where the only habitat available is temporary puddles. The M form larvae develop more slowly but are better at avoiding predators, so they do better in areas where rice fields offer a longer-lasting but more predator-rich habitat. Thus, 'Larval adaptations to exploit temporary vs. "permanent" freshwater habitats account for discontinuities in distribution of the molecular forms ... and provide evidence for ecological speciation' (Lehmann and Diabate 2008, p. 738; see also Lawniczak et al. 2010).

In order to understand how the separate S form and M form populations are diverging, entomologists must acknowledge what they are diverging *from*; recognizing *A. gambiae* S form and M form as *incipient species* requires recognizing *A. gambiae* as a species. More could be said about the leverage provided by a taxonomy that includes the kind **A. gambiae**, but this suffices to show that it is a natural kind for entomology. It shows both that the taxonomy which recognizes the species can do explanatory and predictive work (satisfying the success clause) and that leaving it out of the taxonomy would come at a systematic cost (satisfying the restriction clause).

The research on forms of *A. gambiae* is not, in this respect, especially remarkable. Even a casual perusal of biology literature will turn up further examples, illustrating the predictive and explanatory value of other specific species taxa. So particular species taxa – many of them, at least – are natural kinds for biology.

B.2 The species category

To note the difference between realism and pluralism, we must distinguish particular species (which are given Latinized names like *Anopheles gambiae* and *Anas platyrhynchos*) from the general category **species**. The former are groups of organisms, while the latter is a kind of group. The question of pluralism is this: Does taxonomic success require grouping those particular species together into the kind **species**? Is the **species** category a natural kind?

Two distinct kinds of pluralism motivate answering *no* to both questions. The first is *species concept pluralism*, which results from the diversity of different approaches to conceiving of and identifying species. The second is *species rank pluralism*, which results from the fact that specific species concepts can still fail to distinguish how large or small species taxa should be.

I will first describe each kind of pluralism and explain how they undercut the claim that **species** is a natural kind. Then I consider a response which claims that the diversity of species concepts conceals a single, core concept underneath; I argue that the response fails.

Species concept pluralism

Each species concept offers a different way of divvying up organisms into species, and even if all of the species taxa identified in these disparate ways are natural kinds there may be no interesting kind that includes them all. Rosters of competing concepts are common (e.g. Wilkins 2009, pp. 197–205), but it will suffice to survey them in broad strokes.

The traditional approach to species is to discriminate them in terms of their morphological features. These are the features that dead specimens can be observed to have, when under a magnifying glass at the museum. With developments in molecular chemistry, diagnosis of a species can be made instead in terms of chemical or molecular features. Regardless of the factors that are considered, these approaches share a central commitment: Species boundaries are drawn according to distinguishing features of individual organisms rather than by features of populations or ecosystems. Such features are called *characters*, in contrast with *traits*, so we can call these *character-based approaches* to species.

Applied naively, character-based approaches provide a natural history without any *history*. They construe species membership in terms of present similarity rather than genetic relatedness. After Darwin, we recognize that mere resemblance is not enough. Descent matters. The requirement of connection can also be motivated without direct consideration of evolution. If species are natural kinds, then species boundaries are constrained by the explanatory and predictive value of identifying species. If two organisms were only similar in some respects by chance, there would be no reason to expect them to continue to be similar or to be similar in other respects. There must be *something* in the world that maintains the similarity, and in the typical case this something is an array of biological and ecological processes. I will return to this point later (in Chapter 6), but for now it suffices to note that purely character-based species would not be natural kinds. So the morphological species concept and its relatives provide no argument for pluralism.

Of course, some creatures are sufficiently exotic that contemporary biologists have no choice but to identify species based on primarily morphological criteria. Yet they do not think of these morphological features as *constituting* the species. For example, deep-sea anglerfish are very difficult to observe in their natural habitat, and so their species are divvied up by their distinguishing characters. Theodore Pietsch, who provides an exhaustive catalog of known anglerfish varieties, describes distinguishing characters to provide evidence of common descent; 'a "diagnosis", to use the scientific term' (2009, p. 51).

Yet pluralism does not require that *every* species concept be legitimate. Rather, it just requires that there be more than one. There are at least three broad approaches which are used.

Interbreeding approaches define species membership by looking at the structure of populations. The biological species concept, formulated and championed by Ernst Mayr, takes a species to be a reproductively isolated, interbreeding group. A variant looks instead at mate recognition and whether or not species members *would* interbreed *if* put in proximity to one another.

Ecological approaches look instead at the external factors which enforce group structure. As Marc Ereshefsky explains, according to ecological approaches 'each species occupies its own distinctive adaptive zone, or niche, and the distinct set of selection forces in each zone is responsible for the maintenance of species as separate taxonomic units' (1992, p. 673).

Phylogenetic approaches identify species as the smallest groups that could be subject to evolution and natural selection. A group which includes all of the descendants of a determinate common ancestor is said to be *monophyletic*; a monophyletic group is called a *clade*. Information about clades can be summarized in a *cladogram*, a diagram that summarizes phylogentic relations (see Figure 3.4).

Focusing on clades does not fully specify the category **species**, however, because there will be a larger groups such as genera and families which share common descent in terms of earlier ancestors. There will also be narrower ones. A nuclear family is a monophyletic group, but no biologist would suggest that the household in which I grew up – my mom, my dad, my brother, and I – constituted a species. Species are the level at which natural selection occurs. So a species must be a diagnosable, geographically constrained clade. Diagnosis here is a technical notion, and it means that there must be some characters which distinguish the group. So, for the cladist, character-based approaches return. They are not taken to define species *tout court* but instead to separate the clades which are species from those at other levels.

For some organisms, these approaches will lead to different judgments about which species there are. Since there is a theoretical rationale for each of the approaches, and since each is used productively by working biologists, there does not seem to be a univocal **species** category. If it is such a hodgepodge, then it is not a natural kind.

One might try to deflect pluralism and defend monism by looking at the work of some specific biologists who employ a specific approach to species. Their ability to sustain a research program,

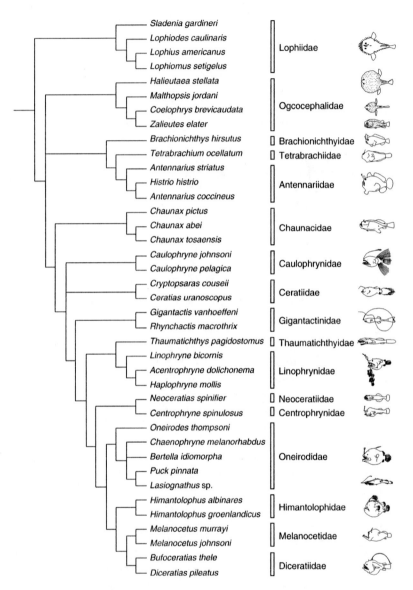

Figure 3.4 A cladogram of anglerfish. Terminal branches (labeled in italics) are species. The bands and pictures on the right correspond to families, each of which is a monophyletic group of species. (© 2010 Miya, Pietsch, Orr, Arnold, Satoh, Shedlock, Ho, Shimazaki, Yabe, and Nishida (2010, p. 16), used under the Creative Commons Attribution 2.0 license.)

one might think, provides evidence that they are applying the *true* species concept. Yet the point of pluralism is not to deny that there could be such biologists. On the contrary, the pluralist insists that each worthwhile concept of **species** could be used to guide an open-ended research program. So, picking a specific concept, we can certainly find biologists who hew to it. For any such biologist, however, there are other biologists working on different open-ended research projects who cannot make use of that same **species** concept. Species concept pluralism, as a philosophical claim about biology in general, is entirely compatible with every individual biologist being able to find a **species** concept that is adequate for their own work – it is simply incompatible with *all* biologists being able to use *the very same* **species** concept.

Species rank pluralism

The hierarchy of different categories is a feature of the Linnaean system of classification that is preserved in biological nomenclature. Species names are given in the binomial format (schematically, *Genus species*), and subspecies in a trinomial format. For example, domestic dogs are *Canis lupus familiaris*, a subspecies of gray wolf *Canis lupus*. Outside of Jack London novels, domestic dogs rarely interbreed with wolves. For most dog varieties, it is a simple matter to distinguish dogs from wolves. So, even accepting a specific species concept, one could argue that dogs should constitute a distinct species. The species concepts themselves are insufficient to determine whether or not they do.

One might worry that the case of dogs is peculiar, because selective breeding is responsible for much of their considerable variety. So consider instead the mosquito species *A. gambiae*, which I discussed a few pages back. Although it is now considered to be one of seven sister species, it used to be that the entire complex was considered a single species and called *A. gambiae*. The complex is thus sometimes called *A. gambiae sensu lato* (meaning: in the broad sense), making the species *A. gambiae sensu stricto* (in the narrow sense). The S form and M form are considered to be *incipient species*, but a case can be made for their already being distinct species. Beyond the actual practice of the community, there is no clear answer as to which of these should be the species level. This is compatible with saying that these taxa are all explanatorily important, and so also compatible with saying that they are natural kinds. It does suggest that **species** is not a natural kind, however, because there is no strong scientific advantage to putting the label 'species' at one of these levels rather than at another.

Species rank pluralism is compounded by species concept pluralism. Consider an example: Between 1982 and 2007, the number of recognized lemur species in Madagascar jumped from 36 to 83. Ian Tattersall argues that, although this is due in part to new discoveries, 'it is overwhelmingly the promotion or re-promotion of subspecies to species, or the recognition of new cryptic species among the nocturnal forms, that has led to the recent enormous increase in apparent lemur diversity' (2007, p. 12). The recognition of all these 'new species' is due both to a change in the species concept employed and the way that they are employed. There has been 'a shift toward fundamentalist notions of the phylogenetic species' in a way that recognizes any 'diagnosable units' as distinct species (Tattersall 2007, p. 21). In the case of lemurs, Tatersall argues that recognizing all of these groups as species comes at an explanatory cost. Distinctions are introduced where there is no scientifically significant difference. We might put the point this way: Not every diagnosable unit of lemurs will count as a natural kind for primatology. The fact that they are carried in under the general banner 'species' hides the deeper question about which taxa *matter*.

Figleaf monism

There is a general response to species pluralism which attempts to recover a core species concept from the various alternatives. It does not hope to secure monism by eliminating all but one species concept. Rather, it aims for a kind of contextualist monism, unity in the midst of difference. Richard Mayden (1997), Kevin de Queiroz (1999), and Richard Richards (2010) all pursue strategies of this kind.

Their idea is that there is a single *theoretical species concept*, according to which a species is a lineage. De Queiroz calls this the *general lineage concept*. The theoretical concept alone is insufficient to identify any actual species. More needs to be said before there is a determinate answer as to which lineages will count as species. The other approaches provide *operational species concepts* which allow the theoretical concept to be applied to real cases. For example, the biological species concept provides a precise criterion which is applicable to some lineages of sexually reproducing organisms. Ecological species concepts provide other criteria. In each case, they offer operational specifications of the same theoretical concept. (The language of 'theoretical' and 'operational' concepts is due to Mayden (1999) and echoed by Richards (2010, ch. 5). De Queiroz expresses it as the distinction between a 'species definition' and a 'species criterion' (1999, p. 80, fn. 1).)

Thus, de Queiroz insists that the only necessary feature of a species is that it be a lineage. He takes features associated with more specific species concepts to be merely contingent; that is, 'properties such as distinguishability, ecological distinctness, and reproductive isolation (to mention a few) are contingent rather than necessary properties of species' (de Queiroz 1999, p. 75). The fact that these may vary independently shows only 'that a single species can belong simultaneously to several subcategories of the general category species. For example, a species can simultaneously be phenetically distinguishable, ecologically distinct, and extrinsically isolated from other species' (1999, pp. 75–6). This makes sense of cases for which differing approaches ultimately lead to the same judgment. For example, Tan et al. (2010) marshal criteria from several species concepts to settle the status of a species of Sepsid fly. Nevertheless, cases like this are not at all what inspired pluralism. The troubling cases are precisely the ones where species concepts lead to different judgments about the boundaries and quantity of species. As Marc Ereshefsky notes, 'lineages that are considered species in one approach often fail to be species in the other' (1999, p. 292). Ereshefsky acknowledges that different approaches all treat species as lineages; he writes, for example, that 'three general approaches to species (the interbreeding, ecological, and phylogenetic) are diverse, but they all assume that species are lineages' (1992, p. 674). Yet, at the same time, he denies that there is any unified conception of species. His position is consistent because the general lineage conception is so remarkably thin.

Consider an analogy with **art** as a general category. There are numerous incompatible conceptions of art, relying variously on conditions of representation, expression, craft, and institutional context. All or most of them would lead us to say that Van Gogh's *Starry Night* is an artwork. The troubling cases are ones in which different conceptions would lead to different categorizations: Andy Warhol's *Brillo Boxes*, pure music, Amish quilting, and so on. One can take this as a reason to be a pluralist about art concepts. (In work with Christy Mag Uidhir (2011), I discuss art concept pluralism at greater length. See also my 2012.) It would be little rebuttal to pluralism to say that all of the various art concepts agree on a minimal necessary condition, namely that an artwork must be an artifact. It is true that all of the art concepts do share this minimal requirement, but it would be perverse to think that the unifying theoretical art concept was just the notion of an artifact and to insist further that institutional and expressive theories of art were mere operational concepts that we use for applying the theoretical concept. The parallel reply to species concept pluralism is not any more satisfactory.

It is a philosophical truism that, given sufficient effort, any *pluralism* can be reconstrued as *monism with parameters*. Yet the reconstrual is not always valuable or informative. Even if there is some value in the thin residue of the general lineage concept, it does not topple the pluralist claim that species are a heterogeneous bunch.

B.3 How species are and are not natural kinds

Pluralism about species is sometimes alleged to foster fallacies and methodological anarchy. For example, Michael Ghiselin suggests that 'pluralism' means that 'species or any other term in our vocabulary can have one meaning in the major premise, another in the minor, and a third in the conclusion' (2009, p. 257). Yet the **species** category, if it is not a natural kind, is not the sort of thing that should figure in syllogisms. Many biologists, recognizing that the category is so diffuse, urge caution. For example, Isaac et al. suggest that 'users must acknowledge the limitations of taxonomic species and avoid unrealistic expectations of species lists' (2004, p. 464). This is especially true when the species concept used affects the *number* of species identified in a domain. Laamanen et al. conclude that 'the number of recognized species depends on the choice of species concept and ... new species descriptions should always specify which species concept was used' (2003, p. 134). Similarly, Tan et al. 'urge all authors who propose new species to be explicit about which species concept they use' (2008, p. 689).

Since the category **species** is not a natural kind, the fate of the word 'species' is just a matter of how we intend to use words. John Dupré suggests retaining 'species' talk in order to provide 'a lingua franca within which evolutionists, economists, morphologists, gardeners, wildflower enthusiasts, foresters, and so on can reliably communicate with each other' (1999, p. 18). Responsibility for identifying species taxa would then be a matter of work within more specific domains. 'Approaches to classification will vary from one group of organisms to another,' Dupré suggests, 'and we should allow experts in particular phyla to decide the most appropriate way of classifying a particular domain' (2002, p. 88).

Another possibility, advocated by Marc Ereshefsky (1999) and Brent Mishler (1999), is to abandon the Linnaean system and 'species' talk altogether. Such an austere approach might come at the cost of making it more difficult to talk about biological taxa that have already been identified and named. Since many specific species are natural kinds, as are many taxa at higher and lower levels, we will still want to be able to distinguish them. An alternate system would need to be devised.

The verbal difference between retaining 'species' as a word or retooling biological nomenclature would have practical consequences. Following them out would carry us beyond the question here, which was whether the species and the **species** category are natural kinds. To summarize: Yes and no, respectively.

Regarding particular species, we saw the example of the mosquito species *A. gambiae*. Recognizing it is important to making sense of the spread of malaria and the ecological differentiation of two emerging subspecies. In this case, the taxon is a real feature of the world and a natural kind. Similar examples could be provided to show that other taxa, too, are natural kinds.

Regarding the species category, concept and rank pluralism undercut the privilege of a preferred *species* level. Although it is important to distinguish *A. gambiae* sensu lato, *A. gambiae* sensu stricto, and *A. gambiae* M form, there is nothing much at stake in arguing which of these three is really, truly the *species*. So **species** is not a natural kind.

I have not dealt here with the third aspect of the species problem: individualism. It is a fundamentally different kind of issue than realism and pluralism. Those two were about whether species taxa or **species** are features of the world. Individualism is instead about *which* features of the world they are. In Chapter 6, I consider the claim that natural kinds are *homeostatic property clusters* (HPCs). An HPC is a causally sustained structure of property regularities, and species are given as paradigmatic examples of HPCs. So the HPC view, like individualism, is a claim about what species *are*. So there is a *prima facie* disagreement between individualism and the HPC view. The alleged opposition is more heat than light, I argue, but that will have to wait for Chapter 6 § D.

C. Thinking outside the box

A traditional conception of *cognition* treats it as something that happens inside the brains of creatures like human beings. More distal parts of the body like hands and feet enact instructions from the brain and return sensory signals. Perhaps cognition extends beyond the brain to encompass other parts of the central nervous system, but according to the traditional view it certainly all happens inside the body. Cognitive phenomena are activities carried out by the organic substrate of nerves. The thought that computers or robots might be capable of cognition is not a direct challenge to the traditional view. Silicon chips and wire replace the organic substrate, but cognition (on this account) is still something that happens on the circuit board. The cognitive activity of a robot, like R2-D2 from *Star Wars*, is contained in its chassis much in the way that my cognitive activities are contained in my body.

The idea of *distributed cognition* challenges the traditional notion, insisting that creatures like us participate in cognitive activities that reach outside the boundaries of our skins. Adherents of the traditional conception see such talk as, at best, picturesque. We do say things like 'She keeps her thoughts in a notebook', but it would be obscurantism to pretend that there is no ontological difference between the thoughts in her head and the ones in the notebook. Ink on paper and chemical patterns in her nervous system are very different things.

Is 'distributed cognition' a natural kind? Or is it just a poetic affectation?

C.1 Tasks and processes

It is typical in cognitive science to distinguish between levels of description or analysis. A much-cited formulation is due to David Marr (1982, pp. 24–7) who distinguishes the level of *computation*, the level of *algorithm*, and the level of *implementation*. At the first level – computation – the system is characterized in terms of what it is expected to do. At the second level, the inputs and outputs are represented in a specific way and connected by a specific algorithm. At the third level – implementation – the algorithm is physically realized in some system. (There has been philosophical debate about exactly how Marr's work should be interpreted, but I make no claim about what Marr himself meant; cf. Wilson 2004, ch. 7.)

Ron McClamrock (1991; 1995, § 1.3) argues that a two-level distinction suffices to do the work of Marr's three levels. He labels the two levels the *task* and the *process*. The computational level specifies the task as 'an abstract and idealized specification of the behavior to be achieved' (McClamrock 1995, p. 20). The lower levels specify the process for accomplishing the task. McClamrock argues that the task/process distinction is a common thread that connects Marr's levels with parallel distinctions made by Noam Chomsky, Herbert Simon, and Zenon Pylyshyn.

To take a simple example, we can ask how a cognitive system like *me* can generate sentences like *these*. The task is a combination of linguistic and motor performance. The process is a complex pattern of neurolinguistic activity in my brain. If we extend the task to include not just generating but also typing the sentences, then the process will also include coordinated muscular activity.

Based on the last several decades of cognitive science, it is plausible to think that the task/process distinction has underwritten successful research and that it has been relatively indispensable. So **cognitive task** and **cognitive process** would count as natural kinds for cognitive science. Moreover, specific tasks like **memory** and specific processes which constitute the mechanisms of memory will also count as natural kinds. One might worry that cognitive tasks like remembering are not things posited by cognitive science, but instead that they constitute the very domain of phenomena which cognitive science aims to describe. In that case, acts of memory might form part of the domain specification rather than a natural kind for cognitive science. This would be fine for what I have to say below; all that I require, for my purposes, is that there be some class of tasks that can be identified as the *cognitive tasks*.

C.2 Distributed cognition

In order to settle whether research in distributed cognition identifies any natural kinds, we need to consider whether its categories allow for prediction and explanation (the success clause) which would not have been possible without them (the restriction clause). Some quotidian cases fail to do so. Take a simple example of someone telling me their phone number so that I can call them in an hour. I might simply memorize the number and recall it later. Alternately, I might write the number on a slip of paper, put the paper in my pocket, and read it off in an hour when I call them. An abstract description of what information needs to be available at what time – of the inputs and the outputs – will treat these as two different implementations of the same task. Yet, one might insist, all of the cognitively interesting activities are done by *me* in either case. The way marks are preserved on paper is straightforward. Treating either process as simply instances of the kind **memory** would elide important differences. A reminder note is just an inscription, but it is at best a weak metaphor to think of memory in the brain as neural inscription. The first process requires me to remember some *particular* sequence of numbers, whereas the second requires me to deploy my *general* skill of writing and reading numerals. The latter skill can be described in terms of some neural implementation. Remembering numbers and remembering how to read are the interesting cognitive phenomena in this case, so there is nothing in this example that requires us to push *cognition* out into the world.

Alternately, consider a system that does arithmetic multiplication. The task defines normative answers for all possible input pairs of natural numbers. The process involves representing the inputs and outputs in decimal and using the usual, grade school algorithm. This process is typically implemented by a person with a pencil and scratch paper. I cannot, without pencil and paper, calculate 372×821. Using the algorithm, I can multiply 372 by 8, record the result, multiply 372 by 2, record the result at an offset from the first, and so on until I can add up the result. This breaks the problem into three less daunting problems. Each subproblem is further broken down by the algorithm; 372×8 involves multiplying 2 by 8, carrying a 1, multiplying 7 by 8, and so on. The algorithm reduces the problem to a stultifying succession of one-digit multiplication problems, seasoned with addition. The computation, the cognitive activity of multiplying three-digit numbers, is not something that happens in my head. Instead, it is performed by the system that includes me, the pencil, and the paper. This is one of the stock examples of distributed cognition; see e.g. McClelland and Rumelhart (1986, pp. 44–8), Zhang and Norman (1995), and Giere (2002, p. 288).

In the example of multiplication, the *distribution* is non-trivial. Only in the simplest cases, such as 3×6, is multiplication like recalling a phone number. Whether I calculate the product with pencil and paper (as I must for 372×821) or in my head (as I might for 4×32), making a calculation requires deploying a general skill. Importantly, the algorithm which I employ when making the longhand calculation reduces the work done in my brain to many smaller matters of skill and memory. Considering merely this succession of simpler operations would leave out what unifies them. Explaining why these disparate operations can occur together to yield the result requires considering the algorithm, and the algorithm is not instantiated entirely in my head. Focusing exclusively on neurobiology would miss the unifying explanation for the parade of small operations. Exploiting the environment allows a brain to do more than it could do in isolation.

Admittedly, the distribution occurs only at the level of implementation. Nothing *in principle* precludes the algorithm being performed without pencil and paper, entirely in a person's head. In fact, I can sometimes calculate such products without paper and pencil – but only on a good day. The thing I have to do is to imagine the paper and hold the entire calculation in my attention. I can only do so, however, because I learned to do them on paper. So the past distributed implementation explains my present brain power. Moreover, this whole operation is only made possible because of a suitable system for representing numbers. The usual multiplication algorithm would be impossible with hash marks or roman numerals. As Jiajie Zhang and Donald Norman show, the cognitive advantages of the standard multiplication algorithm result partly from the structure of Arabic numerals (1995, pp. 284–9). So explaining arithmetic requires considering the symbol system we use. The symbol system allows for inscriptions on paper, and so opens the possibility for the process to extend beyond the nervous system.

The task/process distinction provides a general formula for functional explanation, and so it applies outside the domain of cognitive science. Yet some tasks are ones that are agreed by everyone to be cognitive: calculation, for example. We can say that an activity counts as distributed cognition (d-cog) when a process which implements a cognitive task extends beyond the boundaries of the individual. More precisely, an activity is *d-cog* if (1) the task is such that it would count as cognition if it were carried out entirely in a single mind or brain, and (2) the process by which the task is carried out is not enclosed within the boundary of a single organism. (This account of d-cog appeared in my 2007, although I did not there relate it to natural kinds.)

The example of arithmetic gives us *prima facie* reason to think that **d-cog** is a natural kind for cognitive science. This is not to deny that neurobiological kinds are also natural kinds for cognitive science. Rather, understanding ways in which cognitive agents exploit their environments will require both an appreciation of the larger task (implemented in a d-cog way) and an understanding of the smaller operations (implemented neurobiologically). In the simple case of remembering a number, it seemed as if the latter alone would do. More complex cases, such as arithmetic, require both.

To take another example, some varieties of *transactive memory* will count as d-cog. Daniel Wegner and collaborators (Wegner, Giuliano, and Hertel 1985; Wegner 1986) describe cases in which an intimate couple is capable of storing and recalling information that neither could recall separately. For people who have been in a long-term relationship, this is probably a familiar phenomenon. In some instances, this may best be treated as two feats of individual cognition – two individuals performing distinct but related cognitive tasks. Yet in other cases documented by Wegner et al., such an atomic description is inadequate; the best description treats the couple as implementing a single, unified cognitive task. This seems most apt when one of the two individuals simply would not be able to recover the memory without the other. There is some neurobiological basis for the memory inside the partners' heads, but the process of actually recalling it requires the group effort of cooperative cuing.

Double-blind trials may also profitably be seen as d-cog. We can imagine a single individual testing the efficacy of a drug by administering it to several patients and watching the progress of both those patients and a control group. Even though administering the drug must happen out in the world, all of the cognitive business would go on inside the single observer. Yet we know that actual agents' unconscious biases can subvert such an experiment. So we carry out actual trials using a community of agents, each of whom is only given partial information. The task is to discover the effects of the drug, and the process involves protocols, doctors, nurses, patients, and numbered vials. We can describe a double-blind trial just in terms of what a specific agent knows and does, giving the full account of their neurobiological activity over the course of the experiment. We might go on to do that for all of the people involved. That would miss what is centrally important about the methodology. Moreover, most experimental protocols only describe what happens at the level of the distributed activity. What goes on inside brains is not the issue.

The idea is not that the kind **d-cog** somehow eclipses the kind **cognition in a brain**. Rather, the former captures something which the latter would leave out. As Ronald Giere writes, the framework of distributed cognition 'provides a rich understanding of what is going on. In particular, we can now say that aspects of the situation that seemed only social are also cognitive. That is, aspects of the social organization determine how the cognition gets distributed' (2006, p. 114). Thinking about distributed cognition complements thinking about individual cognition, so recognizing one as a natural kind should not lock out the other. Understanding some important cognitive achievements requires considering both kinds – the tasks implemented inside the person, and the tasks implemented across the person and their environment.

C.3 What the natural kind is not

The considerations of the previous section suggest that **d-cog** may be a natural kind for cognitive science. Inevitably, the case is less decisive than the one I offered for **planet**; cognitive science is not as well developed as astronomy. Before moving on, I offer two caveats. They can be seen as answers to objections which some readers might already be formulating.

First, the criteria which I offered for d-cog provide a sufficient condition for something's being d-cog. Nothing I have said rules out a distributed cognitive process which does not carry out a specific task or which could not even in principle be instantiated inside a single agent. Matt Brown (2011) has pressed this point by noting that we lack a complete analysis of the general kind **cognitive system**. As such, there can be no clear answer as to whether **distributed cognitive system** requires a task specification. Brown suggests that a system which has no precisely specifiable task might still be seen as an *activity system*; e.g. we could come to see ourselves as parts of larger activity systems like labs or academic departments. I am less optimistic that the open-ended activities of such groups will be things that we want to call *cognitive*, since a great deal of non-cognitive work is done by such groups. Regardless, this is a matter to be settled not by metaphysics, but by more cognitive science.

Second, the conception of d-cog which I have discussed is not married to any claims about consciousness. Although some exponents of distributed cognition talk about *extended minds*, the system for conducting a double-blind drug trial is very different than a system that dreams. On my account, insisting that something is a natural kind does not require any commitments regarding its fundamental ontology; so *a fortiori* insisting that **d-cog** is a natural kind of phenomena does not require any commitments regarding its deep metaphysical kinship to **cognition in a brain**.

D. Further examples

The three examples in this chapter are meant to show that the account of natural kinds can be used to throw light on some real cases. The next chapter considers more abstruse questions of metaphysics and realism. Readers interested only in examples are welcome to skip ahead to Chapter 5, where *inter alia* I discuss the examples of baking (§ B.2) and animal signaling (§ C.1), and Chapter 6, where I discuss the examples of polymorphic species (§ A) and water (§ E).

4
Practical and Impractical Ontology

This chapter articulates the connections between my account of natural kinds and several philosophical doctrines. It is common to use the phrase 'practical kind' in opposition to 'natural kind', which reflects a common assumption that the real and the practical are in opposition. The first part of the chapter articulates a view which can reconcile the real with the practical, a view which I'll call pragmatic naturalism (§ A). This talk of reality raises the specter of *realism*, and the second half of the chapter situates my pragmatic naturalism vis-à-vis the landscape of scientific realism. My view of natural kinds fits with what I'll call equity realism (§ B).

A. An unreasonable dichotomy

Robert Martin poses the problem of natural kinds by asking, 'Is the world really objectively divided into real kinds of things, or is it just facts about us (our languages, our cultures, our interests, the ways our minds are set up) that make us tend to divide it in one way rather than in any other?' (2000, p. 215). For the natural kinds we identify, the question he poses is whether they are real or whether they reflect facts about us.

If there were more natural divisions in the world than we could catalog, the divisions that we explicitly acknowledge would inevitably be selected from the range of real kinds. Our cognitive proclivities would, in that sense, determine which way we divide the world. So Martin's dilemma only makes sense presuming that realism about natural kinds requires there to be few of them. In Chapter 1, I called this *the scarcity assumption*. The scarcity assumption fails, but this does not show that natural kinds are ideas or mere fictions. At a freeway interchange, my proclivities are what lead me to take one route rather than another. Yet my taking 787 does not make eastbound I-90 any less real. The *natural kinds we identify* may be real *qua* natural kinds and dependent on us *qua* our identifying.

Nevertheless, many philosophers have assumed the dichotomy that is implicit in Martin's question. Catherine Elgin saddles the realist with the view that 'scientific representations must be *utterly objective*. That is, they must not depend essentially on users or potential users with distinctive interests and points of view' (Elgin 2010, p. 440). Like Martin's false dilemma, this is not something that a realist about natural kinds should accept. Representations *qua* representations exist only because some person or community made them. Scientific papers do not write and publish themselves, and natural languages are products of human culture. Even the staunchest essentialist would not imagine that the word 'water' was a ghost hovering over a lake in prehistoric Europe, waiting to be spoken. The things which are represented may also depend on human action. A realist about dogs and cars should admit without blushing that Weimaraners and 1969 Plymouth Valiants are as they are because of human action; dogs were bred, and cars were manufactured. Yet dogs do not cease to be organisms just because they have been selectively bred, and the kind **dog** does not cease to be a real biological taxon.

In a passage that we encountered earlier (Chapter 2 § D.1), Paul Churchland juxtaposes 'the elect few' natural kinds and 'merely "practical" kinds' (1985*a*, p. 13). In his subsequent argument, which I did not quote before, it is clear that he requires natural kinds to be ones that would appear in a 'unique and final theory' (1985*a*, p. 15). This is an unreasonable demand which presumes the scarcity assumption. As I argued, it is contingent but false. We should not require natural kinds to be uniquely luminous lines etched in a metaphysical fundament, so the distinction between *natural* and *practical* kinds breaks down. Natural kinds support inductive and explanatory success in a domain. Insofar as we care about the domain, acknowledging natural kinds will be useful. This makes them no less real.

In the next several sections, I will address attempts by other philosophers to escape this dichotomy between natural kinds as real features of the world and natural kinds as things we select. To an extent, I am sympathetic to all of them. Considering their positions will hopefully make my own clearer. I begin with a view which is in some ways closest to my own.

A.1 Natural kinds and bicameral legislation

Richard Boyd attempts to capture these two aspects – selected but real – in his account of natural kinds (see *inter alia* Boyd 1999*b*, Boyd 1999*c*, Boyd 2010). In a curious metaphor, Boyd thinks of natural kinds as 'legislative achievements' ratified by a kind of *bicameral legislation* (1999*b*, p. 66). A structure in the world becomes a natural kind in much the same way that a bill becomes a law in the United States, by being approved by both of two legislative houses. One of the houses is human interest and ingenuity. The other house is the causal structure of the world. Since *we* are one of the two houses, it follows from Boyd's view that there were no natural kinds before we humans entered the scene with the particular interests that we have. Even though 'water' as a term was not waiting for us in the world, however, there obviously was water before there were people. So the metaphor of legislation courts paradox. I argue that Boyd's account is not ultimately paradoxical, but that its appearance of paradox points to an important shortcoming. First we need to step behind the metaphor of legislation and consider some of the details of his account.

On Boyd's view, 'the kind **natural kind** is itself a natural kind in the study of the epistemic reliability of human inductive and explanatory practices' (2010, p. 214, my bold). This reflexive move situates **natural kind** as an important part of enquiry *about* science. Since Boyd accepts that natural kinds are enquiry specific, the qualifier is important. The domain of enquiry for which **natural kind** is a natural kind includes the taxonomic work of science. As Boyd sees it, it is enquiry into how 'classificatory practices and their linguistic manifestations help to underwrite the reliability of scientific (and everyday) inductive/explanatory practices' (2010, p. 215).

Boyd thinks of scientific success as a matter of fitting the cognitive and social structure of science to the actual causal structure of the world. He refers to this fitting as *accommodation* to the world's structure. As he sees it, 'the philosophical theory of natural kinds has, as its only subject matter, the ways in which the accommodation demands of various disciplinary matrices are, or could be, satisfied'

(1999c, p. 86). So he defines a natural kind as a category which scientists have used to fit the structure of the world:

> A natural kind is nothing (much) over and above a natural kind term together with its use in the satisfaction of accommodation demands. ... Given that *the* task of the philosophical theory of natural kinds is to explain how classificatory practices contribute to reliable inferences, that's all the establishment of a natural kind could consist in. (1999b, p. 66, emphasis original)

The 'nothing much' in the passage above is to allow for whatever complications are required to handle reference failure and translation of natural kind terms between languages.

I have defended a conception of natural kinds on which they must be able to support inductive and explanatory success. In my idiom, Boyd's idea is this: The domain of enquiry for which **natural kind** is a natural kind includes questions of how and what scientists have discovered. Although it might look like the tautology that natural kinds are of interest to people who are interested in natural kinds, the point is more than that. The kind **natural kind** is a natural kind *for studies of science* and not a natural kind *simpliciter*. It does not have a timeless essence and so will not count as a 'natural kind' according to metaphysically more ham-fisted views. (An analogous point is made by John Dupré 2002, ch. 5.) Moreover, I would add, there is no guarantee that even successful enquiry has identified natural kinds. Recall that my conception of natural kinds requires both that a kind underwrite explanatory and inductive success (the success clause) but also that it be indispensable for doing so (the restriction clause). If the world were so accommodating that most enquiry could succeed using pretty much any taxonomy – if the restriction clause was rarely or never satisfied – then reflections on science would not uncover any natural kinds.

It would be absurd to deny that there was the stuff oxygen long before Lavoisier coined the term 'oxygen'. In an ordinary sense, wherever there is the stuff oxygen there is an instance of the kind **oxygen**. Boyd can agree that the kind applies retrospectively, so that we can now say that there had always been oxygen. Yet he would deny that the kind predated its recognition by scientists. Although there was oxygen before Lavoisier, there was no kind **oxygen**. This is not a point about the metaphysics of kinds, however, since he insists that science is metaphysically innocent. It cannot reach back in time to change the causal structure of stuff in the past. Boyd sees a constitutive connection between natural kinds and natural kind terms, and this is what produces the appearance of paradox.

The crucial issue is that Boyd's account is initially not an account of natural kinds, but an account of *natural kind terms*. He explicitly offers his account as a rival to the 'unbridled enthusiasm for natural kinds and their essences' exhibited by devotees of the Kripke–Putnam theory of reference, and he accepts their unfortunate tendency to approach the problem by way of language (1999*b*, p. 52). Keeping this in mind, Boyd's account makes sense. Natural kinds, as he describes them, are just the counterparts of natural kind terms. Both are defined in terms of the word/world relation. Since there was no term 'oxygen' until Lavoisier coined it, there was no 'oxygen'/**oxygen** relation before Lavoisier.

This might be seen as just a difference in language between Boyd and me. He reserves the label 'natural kind' for categories which have been identified in the course of actual enquiry, and I apply it also to categories which could figure in possible enquiry. It seems possible to translate without loss between Boyd's account and mine, as described in Figure 4.1.

Yet his way of speaking would lead us to odd locutions, such as saying that the natural kind **oxygen** did not exist in ancient Greece.

	Describing a kind that is mentioned in our science:	Describing what it is in the world, apart from whether we mention it:
Boyd's idiom	Oxygen is a natural kind in our chemistry.	If a successful account were to be given of objects and phenomena which comprise the domain of chemistry, oxygen could be a natural kind for that enquiry.
My idiom	We recognize oxygen as a natural kind for chemistry.	Oxygen is a natural kind for the domain of chemistry.

Figure 4.1 Boyd's idiom. Given a particular interpretation of Boyd's account, it is possible to translate without loss between his account and my own. As such, they might just be notational variants.

Of course the ancients had no word for it, I would say, but **oxygen** was around then. It was eventually discovered rather than invented. Although he would agree that the world has its causal structure and that oxygen filled the lungs of people long before the chemical revolution, Boyd has no concise way of talking about what it is in the causal structure of things which makes oxygen suited to be a natural kind. I can say concisely that **oxygen** is a natural kind for the domain of chemistry. Where Boyd would say those words, with his narrower sense of 'natural kind', I would say that *we now know* that **oxygen** is a natural kind for chemistry. If it is a natural kind now, I say, then it was all along. What changed is that chemists learned more about the domain of chemistry than they had known before.

Even accepting the translation, there are two differences between Boyd's approach and mine.

First, he treats natural kinds as relative to the background theories and instruments of enquirers. Although a *domain of enquiry* can be interpreted to include such things, I prefer when possible to construe the domain merely as the part of the world which the enquiry addresses. On Boyd's approach, natural kinds might change when the scientific community's mathematical sophistication increases; on my preferred approach, this does not happen unless the enquiry is one (like sociology of science) which is explicitly concerned with scientists themselves. (Regarding these two senses of 'domain of enquiry', see Chapter 1 § B.11.)

Second, Boyd's account lacks a clear counterpart to the restriction clause. This risks trivializing natural kinds. Consider, as an example, the wise man on Boyowa Island who watches the sky for the appearance of constellations at certain times (Hutchins 2008). These observations allow him to determine the agricultural calendar. In Boyd's language, this accommodates the causal structure of the world: Earth's rotation, the seasons, the life cycle of crops, and so on. The name for a specific constellation thus seems to be, as Boyd would say, a 'term together with its use in the satisfaction of accommodation demands' (Boyd 1999*b*, quoted above). So it seems that the constellations will count as natural kinds on Boyd's account. Still and all, the islander would do just as well identifying *other* constellations. The connection between the observed positions of stars and the seasons is indispensable, but the named groupings of the stars are fungible. So the specific constellations do not satisfy the restriction clause. On my account, they are not natural kinds for the islander's successful astronomy-cum-agronomy. A constellation is generally accepted as exemplary of something which is clearly not a natural kind, so this is a point in favor of my account. (For more about constellations, see Chapter 5 § C.3.)

These difficulties arise because Boyd accepts the agenda of prior approaches which begins with *terms* rather than *things*, with *language* rather than *stuff*. I argued (in Chapter 1 § B.7) that it is a mistake to begin in that way. Nevertheless, Boyd makes a major contribution to thinking about natural kinds. Since scientific success is a matter of fitting our accounts to the causal structure of the world, it follows that natural kinds will typically be held together by causal processes. By looking to causes, we can understand what it is that unifies some otherwise puzzling kinds. This is an important insight that I return to in Chapter 6.

A.2 Amphibolic pragmatism

We are still trying to get clear on how kinds might be at once real and practical. A focus on practice is characteristic of *pragmatism*, so we might look to avowed pragmatists for help.

'Pragmatists,' Richard Rorty writes, 'do not believe that there is a way things really are' (1999, p. 27). Rorty's interpretation of pragmatism is contentious, but it is at least one position we ought to consider. I do not know of any standard label for the view; 'Rortian pragmatism' will not do, since it is not entirely idiosyncratic. I suggest we call it *amphibolic pragmatism*. This follows Mark Wilson, who christens *amphibolic* philosophical views to be ones according to which 'we cannot coherently distinguish between the genuine aspects of the world around us and the personal constructions we happen to bring to their description' (Wilson 2006, p. 77). The amphibolic pragmatist insists that the world is not independent of us.

The view is not a denial of the obvious claim that many parts of the world are beyond our reach. Even Rorty agrees that much of the world is *causally* independent of us. So he agrees that 'if there had been no human beings there would still have been giraffes' yet denies nonetheless 'that giraffes are what they are apart from human need and interests' (Rorty 1999, p. xxvi). Even though giraffes are not causally sustained by our discourse about them, it is we who distinguish them as 'giraffes'. Our language includes a word for them because of our interests. Rorty continues: 'The same goes for words like "organ", "cell", "atom", and so on – the names of the parts out of which giraffes are made, so to speak. All the descriptions we give of things are descriptions suited to our purposes. No sense can be made, we pragmatists argue, of the claim that some of these descriptions pick out "natural kinds" – that they cut nature at the joints' (1999, p. xxvi). This point might be readily deflected by noting that the world is a complex place. There are too many joints in the world for us to have a word for every one of them. Our interests filter *which* kinds appear in our science, but that does not necessarily threaten the kinds' status as features of the world.

This reply will not be enough, however. Rorty insists not only that we might not have had a word for the kind **giraffe**, but that there is no determinate kind there at all. He writes:

> The line between a giraffe and the surrounding air is clear enough if you are a human being interested in hunting for meat. If you are a language-using ant or amoeba, or a space voyager observing us from far above, that line is not so clear, and it is not clear that you would need or have a word for 'giraffe' in your language. More generally, it is not clear that any of the millions of ways of describing the piece of space time occupied by what we call a giraffe is any closer to the way things are in themselves than any of the others. (1999, p. xxvi)

He may be right to say that ant or amoeba-sized scientists would not need a category for giraffe and that alien scientists far off in space would not either. This is only scandalous for natural kinds if **giraffe** is supposed to have some perfectly general *bona fides*. He does not consider the possibility that natural kinds might be domain-specific, so he does not tell us about the enquiries in which **giraffe** might appear. The only interest he mentions is 'hunting for meat', but there is no reason that a vegetarian would be with the amoeba and the aliens in not needing a word for giraffes. The kind **giraffe** is (plausibly, at any rate) a natural kind for various biological and ecological domains of enquiry. The ant, amoeba, and distant aliens would not engage in those enquiries and would be concerned instead with microscopic and astronomical domains. Of course, this does not mean that biology's way of identifying the giraffe is uniquely 'closer to the way things are in themselves' – but that is just to say that

giraffe being a natural kind in that domain does not exhaust what there is to say about giraffe-filled regions of spacetime.

The excess of Rorty's views becomes apparent in this further remark: 'Just as it seems pointless to ask whether a giraffe is really a collection of atoms, or really a collection of actual and possible sensations in human sense organs, or really something else, so the question, "Are we describing it as it really is?" seems one we never need ask. All we need to know is whether some competing description might be more useful for some of our purposes' (1999, p. xxvi). Contra Rorty, it does matter whether a giraffe is just a possible sensation in human sense organs, rather than a long-necked creature galumphing about the savannah. It matters, at the very least, to the giraffe. The veldt extends beyond my *umwelt*. More generally, it matters that we live in a world together with other people and animals. In order for there to be a community – for there to be any interests which count as *ours* and for there to be any account which *we* accept – you others must be more than just a collection of possible sensations in my sense organs. There need not be a single correct account of the world 'as it really is' for that world to be where we live and work.

To strike a more conciliatory tone, I admit Rorty is right that we can believe in giraffes without settling the question of whether they are *really* collections of atoms or clusters of giraffeish tropes. A reductive account is not required to make sense of **giraffe** being a natural kind for biology.

Consider a different example: For purposes of constructing a trebuchet to throw cows, one need not give more than superficial considerations to their internal biology. To put the point that way, however, makes it seem as what matters primarily is *what our interests are* – ranching or siege warfare? Yet the phenomena that these two projects aspire to predict, explain, and control are very different. The domains of biology and ballistics are different, if somewhat overlapping, parts of the world. So it is no surprise that the natural kinds for biology (which might include the cow's diet and metabolism) are different than the natural kinds for ballistics (which would include the cow only as a mass of meat).

And another: In an attempt to show how kinds might be individuated by us – an illustration of amphibolic pragmatism – Sam Page gives the example of **mountain peak**. There is a trivial sense in which the 'number of peaks' depends on the definition of the word 'peak'; if 'peak' meant alligator, then there would be zero 'peaks' in New Hampshire. Even given the topographic presumption that a peak is a highest point rather than a reptile, there is still the question of *which* points count as peaks. The highest point on a bump halfway up a mountain is not itself a peak. Page recounts that

> some hikers insist that in order for a peak in New Hampshire's White Mountains to be a legitimate peak, its summit has to be two hundred feet higher than any saddle connecting it with another legitimate peak. If two peaks are connected by a saddle that is only one hundred feet lower than one of the peaks, then only the higher of the two peaks is legitimate. (2006, p. 332)

So the existence of peaks as objects and **peak** as a kind, he insists, are 'a function of our conventions of individuation' (2006, p. 332).

Notice that this has many of the same features as the *problem of the many* – a sceptical worry first posed by Peter Unger (1980) (Weatherson 2009). Consider a giraffe again for a moment. It has no precise boundary in the world. A molecule at its periphery is not required for its being a giraffe, so there is a giraffe there *excluding* the molecule. Yet adding the molecule does not rule out its being a giraffe, so there is a giraffe there *including* the molecule. These are two giraffes, distinct because one contains the molecule and the other does not. The same argument can be applied all around the periphery of the giraffe, and so (the argument goes) there are indefinitely many giraffes in any place where we think there is a giraffe. This paradox is supposed to undermine ordinary claims of knowledge about giraffes and *mutatis mutandis* pretty much everything else. The problem of the many may be a useful lens for focusing issues in mereology or semantics, but it would do too much for the amphibolic pragmatist. Page allows that the hiker can distinguish a mountain as one peak and not indefinitely many, Rorty allows that the hunter can distinguish the giraffe as one meaty animal and not many, and so the amphibolic pragmatist must think that the problem of the many can be resolved by the specification of the hiker's or the hunter's interests. The skeptical upshot of the problem of the many would make even this discrimination impossible.

Moreover, Page declines to tell us about the interests of the hikers. Suppose that they distinguish peaks so as to plan their weekend trips into the mountains. They want, as many hikers do, to visit 'all the peaks' in their state or in the mountain range. Their enquiry is to identify the peaks on a map so as to climb them all. Since they are planning their trips just at the same time they are identifying peaks, however, the world does not put a strong constraint on how the hikers parse the mountains. The category **mountain peak** is settled by how much climbing the hikers want to do, which is something they are deciding at more or less the same time that they are developing criteria for what counts as a peak. Fungible kinds like these are not natural kinds, because they fail to satisfy the restriction clause; see Chapter 2 § D.2. Perhaps, contrary to my supposition, there is some strong prior constraint on the hiker's enquiry. Then whatever reasons lead them to propose this definition of 'peak' might also make **peak** a natural kind. What needs to be spelled out is what features this definition of 'peak', rather than some other definition, is meant to capture.

A.3 The pragmatists' hope of convergence

Many pragmatists of the non-amphibolic variety parse our contribution to science from the world's by insisting that the *real* is identical to what responsible science will ultimately uncover. Our particular situations and interests determine which things we discover now, but anything that is real is something we will get around to eventually. Even if humanity destroys itself, scientists of some later species might make the discoveries we had not gotten to yet. And even if the whole universe collapses, anything real is something that scientists would have gotten to if they were given indefinite amounts of time. The real world is whatever ideal scientists would talk about in the limit of infinite enquiry.

This idea was originally expressed by Charles Sander Peirce, who writes, 'The opinion which is fated to be ultimately agreed to by all who investigate, is what we mean by the truth, and the object represented is the real' (1992 [1878]), p. 139). The idea is that responsibly conducted science will ultimately yield a true account, not because science is *de facto* reliable but because the ultimate outcome of science provides the criterion for what counts as true. Hilary Putnam alternately calls this position *pragmatic realism* and *internal realism*.

Internal realism suffers from three important failings.

First, science might itself give us reason to think that there are unknowable things. As a matter of general principle, internal realism rules out the possibility of genuinely unknowable facts. So the substantive scientific question of whether or not there are such facts could not even be posed. Yet, our best account of spacetime entails that there are structural facts which – if they were true – could never be shown to be true. John Manchak, extending work by David Malament (1977), concludes that 'general relativity is the sort of physical theory that allows for a wide variety of cosmological models but that, due to structure *internal to the theory itself*, does not allow us to determine which of these models best represents our universe' (Manchak 2009, p. 53). Even if Manchak is wrong, taking relativity seriously requires that we consider the existence of such *unknowables* to be a substantive question – and internal realism cannot do that. (I have written more elsewhere regarding Peirce on unknowability. See my 2005*b*.)

Second, the thought that anything real must eventually be uncoverable by science presumes that the world is simple enough to be exhausted by scientific enquiry. As Paul Churchland points out, the world might be more complicated than that. He writes, 'Just as there is no largest positive integer, it may be that there is no best theory. It may be that, for any theory whatsoever, there is always an even better theory, and so *ad infinitum*' (1985*b*, p. 46). Churchland offers the analogy between theories and whole numbers, but it is not hard to imagine a world even less hospitable to discovery. Let theories be like integers (both positive and negative) and suppose that what a community could discover *next* is constrained by what has been discovered *previously*. Given infinite time, a community that heads off in one direction might discover infinitely many things (corresponding to all the positive integers) but not have discovered everything (the negative integers). Let theories be more numerous, and we can imagine a community discovering infinitely many things (corresponding to all the integers) while failing to discover a higher-order infinity of other things (the non-integer real numbers). These analogies between discoverable features of the world and numbers are rather abstract. They do serve to illustrate some possibilities, however. The world might be complex enough that a scientific community could explore forever and still fail to learn everything about it. Contra internal realism, this is at least a coherent possibility.

Third, according to internal realism, the ultimate science or science in-the-limit defines what exists. A consequence of this is that that science is a science of *everything*, a total science. As I have argued, however, total science is a rationalist pipe dream that looks nothing like real science. Because of this implicit commitment to a total ultimate science, internal realism about natural kinds would presume the mistaken *simpliciter assumption* and preclude the possibility that natural kinds are relative to domains of enquiry. (Regarding the simpliciter assumption and enquiry relativity, see Chapter 1 § B.11.)

One might still express internal realism as the *hope* that the world is sufficiently simple for opinion to ultimately converge on a singular truth. I am more optimistic about this some days than others, but a bet on convergence seems like a bad way to ground our conception of science.

A.4 Engaging the world

To avoid the excesses of amphibolic pragmatism, we still need to say something more about the connection between the practical and the real. In this section, I consider the example-driven account of concepts offered by Mark Wilson.

Wilson is hostile to the tradition of 'natural kinds' so-called, complaining about 'a popular school of contemporary philosophy (characterized by their blithe appeals to the world's alleged *natural kinds*) that severely overestimates the degree to which any of us ... are presently prepared to classify the universe's abundance of strange materials adequately' (2006, p. 56). The 'popular school' Wilson attacks is the Kripke–Putnam account of reference, but his complaint is lodged against anyone who wants natural kinds to all have sharp, necessary boundaries that are well-defined even in exotic thought experiments. The conception of natural kinds I have advocated does not have that aspiration. Rather, it is meant to capture the sense in which we do know something about the structure of the world.

In this respect, Wilson is a potential ally. Success in the world, he insists, means that we have *something* right. He writes: 'Few modes of linguistic behavior ... are likely to last long if they do not embody tolerable stretches of substantive word/world coordination, if only in dedicated patches here and there' (2006, pp. 80–1). Taxonomy can actually be fit to the world, and so there can be natural kinds in the modest sense I have elaborated. Yet this does not mean we necessarily know *which* things we have right. Wilson adds, 'Quite commonly, these supportive correlations prove more recondite in their strategic underpinnings than we anticipate' (2006, p. 81).

So Wilson sees success as the primary evidence that science has got something right about the world. Successful science is sustained by *something* in the world, even though the way in which our language fits to the world is not something we automatically or immediately comprehend. Wilson argues that, 'in many cases, the true nature of a predicate's correspondence with the circumstances it addresses may not prove obvious at all and will require dedicated research to unravel' (2006, p. 80). Putting together the fact that we do get the world right at least sometimes with the fact that we cannot tell the limits of what we have gotten right, he adds that 'successful instrumentalities ... always work for *reasons*, even if we often cannot correctly diagnose the nature of these operations until long after we have learned to work profitably with the instruments themselves' (2006, p. 220). This can be understood as a limitation of our ability to discover natural kinds – although importantly 'natural kinds' in my sense rather than in the essentialist sense which Wilson rejects. To adapt a phrase from William James, there is no bell that goes off to let us know when we have discovered a natural kind for some domain or (after we have discovered one) when we are relying on it beyond the domain where it is a natural kind. That we have done either is an empirical question to be answered by appeal to ordinary sorts of evidence, and the determination is always fallible. Moreover, the metaphysical substance of a category is not transparent. Knowing that a category reflects the structure of the world does not give us immediate fundamental insight into that structure (a point to which I return later in the chapter; see § B).

In numerous examples, Wilson emphasizes the limits and locality of the fit between categories and world. 'Such alignments,' he writes, 'are ... prone to slippage as time goes by' (2006, p. 80). One might insist that the kind of slippage he describes makes for very few natural kinds. A better way to think about it, however, is that the domains of enquiry for which we can give adequate accounts are much narrower than we usually suppose. For a narrow domain of phenomena, the alignments of taxonomy to phenomena that Wilson identifies may be natural kinds. When we do not yet know the contours of the domain in which a category is a natural kind, we press it beyond where it fits to the world. The alignment slips. Wilson would insist that this is not typically something that can be anticipated; rather, we learn the limits of the domain only later.

Despite these caveats, Wilson does think that we get the world right. Christopher Pincock calls this 'a "patient" and limited form of scientific realism' according to which 'we can be sure that we are engaging with an objective reality, but we cannot be confident that we have the description of this reality right' (2010, p. 120). He also notes Wilson's denial that there is any general formula for refining the description; as Pincock puts it, 'There is no general test that can assure us that we are getting things right' (2010, p. 121).

Wilson's talk of 'instrumentalities' nicely fits with the idea that kinds are a practical matter, while still allowing for a modest, fallibilist realism. The insistence that knowledge is to be worked out by domain-specific methods rather than by some general test for truth also fits with my own account of natural kinds as domain-specific. Although he would perhaps bristle at the suggestion, many of the cases he studies can be seen as examples of how natural kinds (in my sense) have been discovered.

A.5 The practical as leverage on the real

I have tried to argue that natural kinds, on my account, can be both practical and real. The dialectic, which is summarized in Figure 4.2, can be understood in this way:

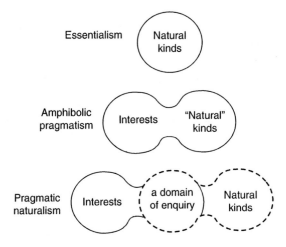

Figure 4.2 Three views about natural kinds. The first two privilege natural kinds as *real* or *practical* (respectively), but the better view can see them as both.

The view that natural kinds must be real and non-practical, exemplified by traditional essentialism, holds that the world's stock of natural kinds forms a single list. Although we can study the ones that interest us, the existence of each *qua* natural kind is a feature of it in isolation. They are like the items in an omnibus catalog; even if we cannot order everything in the catalog, there is only one registry for what is available. The view fails because the world is too complex to be disinterestedly tallied up, because intrinsic properties are insufficient to characterize many natural kinds, and because natural kinds do not have their status as a flat-footed matter independent of any context. (This failure follows from the Chapter 1 arguments against the scarcity, intrinsicness, and simpliciter assumptions.)

A disillusioned essentialist, still insisting that natural kinds must be natural kinds *simpliciter*, might become an amphibolic pragmatist. On that view, we identify the kinds that we want to identify. Our interests determine which kinds seem 'natural' – which is to say that none are genuinely natural.

Since success is only possible because the world will sustain it, however, we do learn about the world. Recognizing that natural kinds are relative to enquiries allows us to recognize that at least some of the kinds we have identified might be natural kinds in the domains where we exploit them. We engage in the enquiries that interest us and investigate the domains that we care about. Within some of those domains, there are natural kinds. Call this *pragmatic naturalism*. (The label 'pragmatic realism' would also be apt, but it is already taken by Putnam as the handle for a different view.)

Which phenomena are to be explained is a matter of which questions we hope to answer. Yet the phenomena are features of the world. In a sense, the domain of enquiry screens off natural kinds from direct influence by our interests.

Pragmatic naturalism applied to natural kinds means that our categories do, in suitably specified enquiries, fit to the structure of the world. The fit is not some general fact. Natural kinds are domain-specific things. We can learn that we have identified natural kinds and we can learn the limits of their proper domains, but the methods by which we do so are the methods of empirical enquiry. The identifications we make are always limited and fallible, because we can always learn more and we might always discover respects in which we had been wrong.

Our interests determine the domains into which we enquire, but there may be natural kinds to be found in them. Whether we discover natural kinds depends both on the domain and on our ingenuity, but whether they are there does not depend on us.

B. Deep metaphysics

There are many distinct issues tagged as 'realism', but it will help to distinguish two.

First, there is a broad metaphysical concern about whether the world as a totality has a mind-independent existence. Anti-realists, opposed to this general kind of realism, might be idealists or amphibolic pragmatists. A realist in this sense might think that there is no unique true description of the world, but the limitation must arise from the fact that the world is (to put it crudely) too big to exhaustively describe rather than from a claim that the world is a projection of thought. Specific realist accounts typically go on to describe the world as it is, and there are many options; the realists' world might be a collection of entities, substances instantiating universals, a structure of properties, a cloud of tropes, or some other deep metaphysical ontogoria. Call such views *deep realism*.

Second, there is the specific question about whether scientific descriptions of the world should be understood to be at least as real as descriptions of objects that we encounter in daily life. Semantic anti-realists deny this, insisting instead that words like 'electron' should be understood to operate differently than words like 'pain' or 'baguette'. Epistemic anti-realists, such as constructive empiricists, admit that words like 'electron' do refer to unobservable things but deny that we should believe in *electrons* and other unobservables. The realist in this sense is an egalitarian about the posits of science, insisting only that *electrons* are on much the same footing as *pain* and *baguettes*. Call this view *equity realism*.

It may seem as if equity realism just ought to be called *scientific realism*, and I feel some temptation to do so. I hesitate because the label 'scientific realism' is entrenched in a deep swath of literature, and it has many entanglements that go beyond equity realism. Scientific realism is often taken, both by proponents and detractors, to be the view that the claims of our science are true – this is too strong, of course, so the qualified restatement is that most mature scientific theories are probably, approximately true. The equity realist is not committed to any such blanket conception. Rather, the equity realist simply denies that there is any philosophical distinction which wrenches the more theoretical posits of science from the quotidian stuff with which we are more readily familiar. There may be, *contra* the scientific realist, reasons to deny the existence of some entities which appear in mature science – but science might just as well give us reasons to deny the existence of some objects which had been part of our pretheoretic picture of the world. The best account of science might be realist about some things and anti-realist about others, but the difference must be responsive to evidence and scientific considerations. The equity realist insists that there is no single fault line along which to split the real from the unreal, and so *a fortiori* no such line motivated just by philosophical reflection.

The pragmatic naturalism that I articulated in the previous section maintains that natural kinds, though practical, are real. This commits me to equity realism about kinds. **Electron** as a natural kind for physics and chemistry is on the same footing as *electrons*. One might construct a slightly weaker equity claim which would be neutral on the question of scientific realism versus anti-realism. It would insist, as I do, that natural kinds identified in our science are in an important respect on par with the unobservable entities posited by successful science. A constructive empiricist (like Bas van Fraassen 1980) or an instrumentalist (like Kyle Stanford 2006) could agree about that, because they typically agree with scientific realists about which scientific account is the *best*. They disagree with realists only about whether this superiority makes an account worthy of belief. The anti-realists say that the best theory is only to be accepted or used as a predictive device. What they say about entities, they should also say about natural kinds. If such an anti-realist account of science were tenable, then what I have said here in a realist idiom could be reinterpreted in an anti-realist way. I have argued elsewhere that such anti-realism is untenable, at least as a blanket policy, but those concerns are independent of equity about natural kinds. Regarding van Fraassen, numerous problems arise (Magnus 2003, ch. III). Regarding Stanford, I think he is better understood in a way that is not entirely anti-realist (Magnus 2010).

B.1 Bad arguments for realism

Many debates about realism are conducted at a rather high level of abstraction. Arguments are given that we should believe in things like electrons, but the arguments do not depend on any of the particular evidence about electrons. Rather, they appeal to general considerations which are taken to favor realism about the posits of mature, successful science. I call these *wholesale* arguments, because they try to seal the deal for realism in one transaction. The contrast to a wholesale argument is a *retail* argument. A retail argument for the existence of electrons would consider facts peculiar to particle physics or to experiments which establish the existence of electrons. Admittedly, the difference between retail and wholesale arguments is a matter of degree – arguments can rely on differing amounts of particular detail. Nonetheless, I have found the distinction to be useful for separating different arguments and interpretations of arguments (Magnus and Callender 2004; Magnus 2010, 2011b). Debates about scientific realism, in the broad sense I discussed above, provide and address wholesale arguments.

The most common wholesale argument in favor of scientific realism is the No-Miracles Argument (NMA) which holds that the remarkable success of science would be miraculous if our theories were not true; so, it invites us to think that our theories are (probably, approximately) true. Robert Martin suggests that the NMA is the primary argument for realism about natural kinds. He writes, 'The main argument for category realism is this: it's a fact that some categories work – they allow explanation – and some don't work. The only possible explanation for this fact is that the categories that work are really out there – really features of the mind-independent universe' (2000, p. 216). Whatever one might think of the NMA for theories, the NMA for categories is unsound. In domains where success is easy, our taxonomy would work because it is just one of a great many taxonomies that would work. So success can be explained by the weak demands of the domain instead of by the reality of the kinds. In terms of natural kinds as I characterize them, the NMA would require that a kind's merely satisfying the success clause be strong reason to think that it is a natural kind. This leaves out the important role of the restriction clause.

Satisfying the restriction clause is non-trivial. There is no guarantee that kinds which appear in a mature science will necessarily satisfy it. So my conception of natural kinds leaves open the possibility, in the abstract, that there are no natural kinds whatsoever. Showing that there are natural kinds in a domain requires considering the contingent findings of enquiry in that domain. As a result, our best guess about whether and which natural kinds there are in a domain will be upset if our scientific account of that domain must be considerably revised. Although there is some ground to be cleared generally – as in this chapter – the account of natural kinds forces us to supply retail arguments.

In a paper arguing for the correspondence theory of truth, Philip Kitcher attempts to offer a refined version of the NMA. The view he defends is that 'realism endorses a *Success-to-Truth* rule' which underwrites inference from '*S* plays a crucial role in a systematic practice of fine-grained prediction and intervention' to the conclusion that '*S* is approximately true' (2002, p. 348). On the face of it, it may sound as if Kitcher's NMA and my account of natural kinds are saying the same thing. After all, I would agree that a kind is a natural kind if it 'plays a crucial role in a systematic practice of fine-grained prediction and intervention'; i.e. if it is indispensable for inductive and explanatory success in a domain. Nevertheless, there are important differences. Kitcher's *S* is a sentence, and he argues for its truth understood as correspondence. A natural kind is not true – not in the sense of correspondence nor in any other sense – nor is it false. Sentences have truth values, while taxonomies do not. Furthermore, Kitcher's 'approximate truth' is correspondence to the world *simpliciter*. Since a natural kind is enquiry-specific, it will only support success in its domain. To reiterate lessons of § A.4: When we think some part of our taxonomy in a domain is a natural kind, we should think that the kinds correspond to features of that domain. We have learned something about the structure of that domain. Yet it will not necessarily be obvious exactly what the structure is that supports our success. The details will probably require further investigation.

As a corollary to this, the account of natural kinds I have offered will not settle questions of ultimate metaphysics. Natural kinds are features of the world, I argue, but I have no story to tell about the deep metaphysical substance of them. This is not a scandalous gap in my account. Quite the contrary, caution about deep ontology is a counterpart to the focus on retail arguments. Retail arguments for believing in particular things can give us good reasons to believe that those things exist on the basis of their connections to other things, while leaving questions of things' fundamental nature either unmentioned or unresolved. Astronomers are and pretty much always have been realists about planets like Mars and Venus, for example, even though we only learned about their constitution in the twentieth century – and even though there is still a good deal we do not know about them. Jean Perrin's experiments provided convincing reason to believe in atoms, even though our present account of subatomic structure is very different than the theories on offer in his time.

B.2 Realism and metaphysical depth

What kind of deep realist story might we tell on this basis? Retail arguments about the existence of some *things*, like electrons or atoms, fit comfortably with entity realism. Entity realism is a view, associated with Ian Hacking (1983) and Nancy Cartwright (1983), according to which science tells us about the existence of unobservable things but not about their fundamental natures. The fact that we can exploit electrons in a diverse array of contexts shows that there are electrons, even if we lack an exhaustive or perfect description of them. In Hacking's pithy formulation: Where we can spray them, they are real (Hacking 1983, p. 24; Cartwright 1999, p. 34).

Yet the things in which we believe are not electrons or atoms in-themselves, but rather the structure which sustains the connections that provide us evidence. So one might see this as a kind of structural realism. Structural realism is a view that we do not come to know about entities as such but instead about structures. The view was introduced in recent decades by John Worrall (1989), and considerable debate has ensued. It is now typical, following James Ladyman (1998), to distinguish *epistemic* and *ontic* structural realism. Epistemic structural realism holds that we do not know anything about unobservable individual objects themselves, although they do or at least could exist. Ontic structural realism denies that there are individual objects as such.

Ladyman's most developed statement of ontic structural realism is in his work with Don Ross. On their view, what we would ordinarily think of as objects and kinds are all *patterns*. They adapt the notion of real patterns from Dennett (1991) but argue that a pattern is not – as one might naturally think – an arrangement of underlying things. Rather, patterns are the basic elements of their ontology. This means that 'neurons, peptides, gold, and Napoleon are all real patterns', but so are 'quarks, bosons, and the weak force' (Ladyman and Ross 2007, p. 300). Even if there is some sense in which Napoleon is a pattern, it is less clear that we should think of **Napoleon** as a natural kind.

Similarly, the particular electrons sprayed at a target on a specific occasion are not the same as the general kind **electron**. This is not a decisive objection, but it does suggest trying to synthesize entity realism and structural realism.

Anjan Chakravartty attempts to do this with a view he calls *semirealism* which holds that, '[c]oncrete structures are relations between first-order properties of things, so to know them is to know something qualitative about relations' (Chakravartty 2007, p. 41). A particular is a thing with properties and relations, so kinds correspond to clusters of properties and relations that tend to occur together. Chakravartty explains:

> Properties, or property instances, are not the sorts of things that come randomly distributed across space-time. They are systematically 'sociable' in various ways. They 'like' each other's company. The highest degree of sociability is evidenced by essence kinds, where specific sets of properties are always found together. (2007, p. 170)

Beyond what Chakravartty calls 'essence kinds', there are sets of properties that tend to be found together but are not always conjoined; they are loosely sociable. He adopts this metaphor of properties being *sociable* from Bertrand Russell (1948). It is 'of course ... just a metaphor' but it is a metaphor that resists analysis (Chakravartty 2007, p. 170). I take it that, for the semirealist, the association of properties in the fabric of the world is just a basic notion. There are some kinds which simply *are*, with no explanation to be given beyond the brute sociability of their properties.

As a view of natural kinds, semirealism is 'somewhat deflationary'. Chakaravartty writes, 'Kind-talk simply reflects distributions of causal properties. One describes cases in which distributions are sociable enough to be useful, convenient, or interesting as instances of kinds' (2007, pp. 178–9). His idea seems to be that there are indefinitely many ways that properties associate with one another in the world. Our selection of some associations as *natural kinds* must select from those, and Chakravartty thinks that this can only be done on the basis of our interests. Strangely, the semirealist finds common cause with the amphibolic pragmatist. The unholy alliance can be broken in this way: In his discussion of natural kinds, Chakravartty implicitly accepts what I have called the simpliciter assumption and so fails to consider that natural kinds might be constrained by the enquiry for which they are a natural kind. If we take domain relativity seriously, then the success and restriction clauses work to narrow down which kinds count as natural. For the domain of astronomy, **planet** is a natural kind. Of course, our interests and what we find useful lead us to do astronomy – but **planet** would be a natural kind for that domain whether we did astronomy or not.

In maintaining both things and kinds, one might instead offer an inflationary ontology that construes natural kinds as abstract entities. Stephen Mumford writes, 'Natural kinds are best understood as types or universals; hence the problem of kinds is a problem about universals' (2005, p. 433). Whereas Chakaravartty speaks of properties sociably associated, Mumford speaks of 'the world's modes [which] naturally divide into types' (Mumford 2005, p. 433). The disagreement is, perhaps, just the medieval debate between nominalists and realists about universals. Even if the twenty-first-century details change the substance of the debate, it carries us far beyond the special concerns of philosophy of science. Rather than pursue the matter further, it is time to step back from the abyss.

The characterization of natural kinds which I have given does not make claims about fundamental ontology. On the basis of evidence, we form fallible beliefs about natural kinds, and I have suggested some of the status that natural kinds have in the world. The pragmatic naturalism I have advocated is committed to equity realism, and so it is metaphysical in the sense that it goes beyond experience to say something about what the world is like. Yet there are things my account does not say. What natural kinds are *in rerum natura* is a matter for the deep realist to sort out. My account is compatible with several possible outcomes of that debate or even with a principled rejection of the further question.

5
The Menace of Triviality

In this chapter, I consider some objections to the account of natural kinds that I have developed and illustrated in the previous chapters. The objections allege that the account of natural kinds is too liberal. It counts too many kinds as natural kinds, they say, making the modifier 'natural' a mere flourish and robbing it of any scientific or philosophical significance. To put the charge differently: If every kind turns out to be a 'natural kind', then we might just as well say that there are no natural kinds.

I distinguish three different arguments, each offering a different basis for the allegation:

In § A, the objection is that any set of objects will be similar in some respects and so will form a minimally projectible kind. Features of my account, including domain relativity of natural kinds and the restriction clause, suffice to deflect this objection.

In § B, the objection is that contrived domains of enquiry might make anything count as a natural kind. Like John Dupré's promiscuous realism, my view will acknowledge natural kinds for domains like cooking. However, this does not reduce to absurdity. I discuss the specific example of baking and suggest that there is nothing wrong with the idea that baked goods might form natural kinds.

In § C, the objection focuses on domains in which some organisms distinguish classes of things which would not otherwise count as natural kinds. For example, aerial threats to meerkats form a natural kind for meerkat ecology because of the organized response that meerkats give to them. The worry is that this might extend to superstitious humans, who identify dubious categories of things like constellations. It does not generalize so readily.

If 'natural kind' and 'kind' did turn out to be synonyms, I would readily give up the distinction. However, the objections fail to show that natural kinds collapse into triviality. Insofar as my account is more liberal than some others – insofar as I acknowledge more natural kinds – it is just a sign that the world has a richer structure than those other accounts acknowledge.

A. Cheap similarity

Anjan Chakravartty analyzes natural kinds as corresponding to properties which occur together, properties which are 'sociable' in a metaphysical sense. (This is part of Chakravartty's *semirealism*, which I discussed briefly in Chapter 4 § B.2.) He writes:

> Everyday and scientific kinds recognized in the course of systematizing nature are perfectly natural, but what is natural goes well beyond what is useful, convenient, or interesting in everyday and scientific contexts. Nature is composed of distributions of property instances, only some of whose patterns of sociability we consider and investigate. There are, it would seem, unimaginably many natural kinds. (2007, p. 178)

This can easily become an objection. Even a gerrymandered collection of things will have some properties in common; in Chakravartty's idiom, its members will have at least some properties that are at least a little bit sociable. So any arbitrary class can support some inductive generalizations. Thus, every kind ends up being a natural kind.

Some natural kind enthusiasts are less liberal than Chakravartty. T. E. Wilkerson, for example, insists that some predicates simply are not projectible and so do not correspond to natural kinds. He gives the example of **table**, which he insists is not a natural kind. Merely by knowing 'that the object over there is a table,' Wilkerson writes, 'I am in no position to say what it is likely to do next, nor what other things of the same kind are likely to do.' A table might be sturdy or not, depending on whether it consists of rotted wood or stainless steel. He concludes, 'In short, because there are no real essences that make ... tables tables, I cannot even in principle make sound inductive projections about ... tables as such' (1988, p. 31). The modifier 'as such' in the last passage is crucial. Wilkerson acknowledges that one might rely on the steel table's being sturdy, but he thinks that this could only be in virtue of that table's being of the kind **steel thing** rather than its being of the kind **table**. (We encountered Wilkerson's claim earlier, in Chapter 1 § B.8.)

Wilkerson's example simply misdescribes how we relate to things like tables. Consider, for example, the table in the philosophy department lounge. At lunch on a given day, I set a bowl of hot soup on it with the implicit expectation that the table will support it. I do this without reflecting on what material constitutes the table. I do not know – nor do I particularly care – what is underneath its laminated surface. It is a table, and that is all I need to know. So it is with tables in restaurants, copy shops, post offices, and throughout daily life. If I consider sitting on a table, I test it by leaning against or shaking it. Again, the question of what constitutes the table is irrelevant; a metal table might be rickety, and a wooden table might be stable. By knowing that an object is a table, we know a great deal about what it will do next. Insofar as there is uncertainty, because tables admit of variation, the gap cannot typically be filled by information just about the composition of the table. Of course, this does not show that **table** is a natural kind. The example is underspecified, because we have not said what enquiry we are considering. The point is simply that – contra essentialists like Wilkerson – many classes of things are such that we can say (roughly, at least) what a member of the class is likely to do next. All classes support *some* inductions. So the objection remains.

The objection applies most readily to a conception of natural kinds consisting only of what I called (in Chapter 1) the induction assumption, a conception which accepts any minimally projectible predicate as a natural kind. An arbitrary collection of things, because they will inevitably have something in common, is assured to support *some* inductions. Pressing the attack against my conception of natural kinds, though, one might argue that any collection will thus meet the success clause at least to some minimal degree.

To reply: This will not hold for every arbitrary collection, because it does not follow from the mere assurance that there are *some* similarities that the similarities will obtain within the domain of phenomena which the enquiry addresses. Even if they do, other kinds might support the same inductions – in which case the collection fails to meet the restriction clause, and so it fails to be a natural kind for that domain.

Chakravartty's entirely general insistence that 'what is natural goes well beyond what is useful, convenient, or interesting in everyday and scientific contexts' (cited above) presumes that natural kinds are domain independent. Once we specify domains of enquiry, the objection loses much of its force. Consider the example of mosquito species as natural kinds for epidemiology (from Chapter 3 § B.1). They count as natural kinds for that domain not just because the members of each species share some genetic and phenotypic traits – rather, they count as natural kinds because malaria has a significant impact on public health and the mosquito species are central to explanations of malaria's spread. Tuna species, even though they also chart patterns of similarity, are not natural kinds for epidemiology. For a broader biological domain, the study of animals including both insects and fish: both mosquito and tuna species may be natural kinds. In each case, it will depend on the place of that kind within the specified domain.

In the next section, I consider an objection that attends directly to the domain-dependence of natural kinds to argue that peculiarly human domains will license unnatural kinds.

B. Project-relative kinds

Since I have not offered any conditions which a domain must meet in order to count as a domain of enquiry, then we can ask after natural kinds even for domains that seem *unscientific*. If we consider a domain like *stuff in pockets*, the objection maintains, we end up acknowledging 'natural kinds' which are hardly worthy of the name.

This kind of objection is often posed against John Dupré's *promiscuous realism* (Dupré 1993). Dupré argues that the kinds which are of interest to cooks and carpenters are just as real as kinds which are of interest to ecologists – a position which parallels my allowance that domains of cookery or carpentry might have natural kinds. So I start by exploring how promiscuous realism has been misunderstood and maligned.

Turning back to my own account of natural kinds, I consider the example of whether or not there are natural kinds for baking. There are – or at least there might be – but I argue that this is not a demerit of my view.

B.1 Promiscuous realism

Thinking about *species*, John Dupré argues there is no single, privileged way to distinguish categories of living things. Moreover, he argues that classifications for human purposes are just as legitimate as those made according to criteria like interbreeding or common descent. Cooks distinguish broccoli from Brussels sprouts, for example, despite their botanical similarity; biologists classify both as varieties of the species *Brassica oleracea*. Carpenters group pines together on account of their timber, despite their botanical difference. Dupré's examples are garlic and onions (2002, p. 34) and cedars (2002, p. 29). He embraces alternatives like these, insisting that 'there are many sameness relations that serve to distinguish classes of organisms in ways that are relevant to various concerns ... [and] ... none of these relations are privileged' (2002, p. 33). He dubs this view *promiscuous realism*.

It is often dismissed as only an ersatz realism. Critics allege that it reduces to relativism, to anti-realism, to what I called amphibolic pragmatism (in Chapter 4). Such charges abound: Stephen Mumford writes, 'Dupré ... denies essentialism and is so promiscuous about natural kinds as to practically deny them' (2005, p. 423). The editors of a volume on pluralism allege that 'promiscuous realism is hard to distinguish from radical relativism' (Kellert, Longino, and Waters 2006, p. xiii). Marc Ereshefsky complains that 'Dupré's pluralism is too promiscuous. ... It legitimizes taxonomies that are in no way based on scientific reasoning' (1992, p. 687). James Ladyman, Don Ross, and John Collier allege that promiscuous realism 'amounts to abandonment of the metaphysical ambition for a coherent general account of the world' and that it is a 'denial of the possibility of philosophy' (2007, pp. 194, 196). Perhaps most damning, George Reisch (1998) alleges that promiscuous realism would put creationism on the same footing as evolutionary biology. If Reisch were correct, then promiscuous realism would be a Trojan horse, entering the city with the promise of real kinds but unloading methodological anarchy under cover of darkness.

Although Dupré is often misread on this point, he does not endorse accepting *any* category as a legitimate species taxon for biology. Reisch claims that Dupré could only draw a distinction between creationism and biological science if he could 'show that the epistemic interests and efforts of creationists to structure the world are somehow not legitimate or genuine' (1998, p. 341). However, this gets the matter backwards. Dupré need only say that creationist taxonomies fail to successfully chart the domain of scientific biology. Biologists' species taxa – although not an entirely unified lot – do a better job structuring the description and explanation of organisms and populations, their complexity and history, than would any creationist alternative. A *divinely ordained species* concept would have no place in scientific enquiry because it would not sustain any inductive or explanatory success, and so creationism has no place in the science curriculum.

Ereshefsky's more general worry that promiscuous taxonomies 'are in no way based on scientific reasoning' (cited above) similarly misses its mark. The appropriate taxonomy *for biology* is based on scientific reasoning about the domain of biology. Yet there are other domains which overlap the domain of biology, and scientific reasoning may generate different taxonomies for those alternate domains. Dupré is explicit that culinary kinds – though real – are not species. Species are identified within biology, and culinary concerns are not the concerns of biology. There is no place in biology for a *gustatory species* concept which divides up *Brassica oleracea* in a way suited for restaurant menus or that categorizes most animals as close relatives of chickens.

The condemnation by Reisch and Ereshefsky presumes that there is some clear demarcation criterion, a way to divide science from non-science. Yet none of the demarcation criteria that philosophers have offered will suffice to celebrate chemical and biological kinds while banishing cookery kinds. Karl Popper (1963, 1965) famously argues that science is characterized by the critical attitude and that scientific theories must be falsifiable. There are familiar worries about the notion of falsification, but we need not rehearse them. The criterion clearly will not part physics, biology, and other sciences from cautious and empirical cooking. A chef can make predictions on the basis of culinary taxonomy, and the predictions might turn out to be false.

Note, however, that the creationist bogeyman can be vanquished without a binary demarcation criterion. All that is required is a difference in degree between well-confirmed and disconfirmed scientific accounts, between convincing science and junk science (Kitcher 1984/1985). Some of the most successful parts of physics and biology are convincing. Creationist accounts, if treated as science, are unconvincing junk. Whether or not there are well-confirmed culinary theories is then an open question.

The worry expressed by Ladyman et al. is somewhat different. It alleges that Dupré turns his back on *metaphysical ambition*. There are two ways of understanding this worry. One is based on the idea that metaphysics must be fundamental ontology. E. J. Lowe voices a common sentiment when he insists, 'Metaphysics, properly conceived, is the study of the most fundamental structure of reality' (2009, p. 108). Dupré certainly is not doing *that*. However, as I argued in Chapter 4, one can be a realist about natural kinds without claiming to know about their being-as-such – one can be a realist without being a deep realist. An account of natural kinds like Dupré's or like mine can be silent on questions of fundamental metaphysics without insisting that fundamental metaphysics is in principle impossible. A second way of understanding the worry would preclude this rapprochement. The claim would be that the world of promiscuous realism is too Balkanized, and so there could be no unified metaphysical picture in which it fits together. I do not see how this could be sustained, however. There is a coherent world where a plant can be both a member of the kind **Brussels sprouts** and of the species *Brassica oleracea*.

T. E. Wilkerson, himself a champion of essentialism, concedes that promiscuous realism 'is a version of realism, because the similarities and differences that underpin each system of classification are independent of our beliefs and theories, but the realism is promiscuous, because there will be an indefinitely large number of possible systems of classification, each one reflecting a different interest and concern' (1993, p. 5). Wilkerson fairly summarizes Dupré's view in this way:

> [T]he features fastened on by, for example, farmers and gardeners are *real* features of the world. It is a fact about the world, not a fact about us, that some animals produce milk, that some plants survive the British winter, that some plants are edible, and so on. The only relevant fact about us is that if we are farmers, we will concentrate on one set of features; if we are gardeners, we will concentrate on a rather different set; if we are zoo keepers, we will concentrate on another set; and so on. (1993, p. 11)

Robin Hendry argues that pluralism ultimately *underwrites* realism. He explains, 'the recognition that there are many divisions in nature excludes that there are none, and the recognition that a particular discipline might have studied *different* divisions has no tendency to undermine the reality of the divisions it does study' (2010, p. 138). This is also the point of pragmatic naturalism; see Chapter 4.

B.2 Cooking up natural kinds

Dupré argues that there are many different real kinds, which is not exactly something I would say. According to the account of natural kinds I have been defending, the question of natural kinds can only arise given a domain of objects and phenomena. So the natural kinds I defend are not real kinds *simpliciter*, but instead real features of the domain. Nevertheless, I readily allow for domains that include objects like *cookies* and phenomena like *cooking until golden brown*. In this section, I approach the domain of baking by way of the account given by Alton Brown. Brown plausibly identifies natural kinds for baking, and this should be no embarrassment to realists.

In his television show *Good Eats* and in his books, Alton Brown explains the physics and chemistry behind various recipes: what flour does at a molecular level, how butter makes biscuits fluffy, and so on. In his book on baking, Brown claims that 'the greatest analytical tool in the world is classification' (2004, p. 7). He provides a taxonomy of baked goods according to which 'the primary mixing methods ... make baked goods what they are' (2004, p. 8). For example, traditional muffins, soda bread, pancakes, and waffles are all prepared using what Brown calls the *muffin method*. Biscuits, scones, shortcakes, grunts, and pie-crusts are prepared using the *biscuit method*. Importantly, these methods are not merely a matter of preparation. They also correspond to different ways of leavening. In all the muffin method baked goods, air bubbles are generated by baking soda reacting with acid. In biscuit method baked goods, bubbles are generated by small bits of solid fat which melt during baking. Even if a soda bread and a biscuit are similar end products, the chemical processes that generate them are different.

Note that Brown's usage does not perfectly track the way that words like 'muffin' and 'biscuit' are typically used. He acknowledges that 'the accepted method of classification – the one used in most cookbooks – is nomenclature-based: pancakes, biscuits, rolls, and so on. I don't think this is any more a "system" than sorting books by color. Names just don't mean that much' (2004, p. 7). The usual terminology that groups baked goods by their size and shape is rather superficial. It is appropriate to be a nominalist about such an arbitrary system. Something sold as a muffin, Brown says, might instead be 'a cake masquerading as a muffin – better known as a cupcake' (2004, p. 87). This is not mere pedantry. In Brown's system, there is a principled distinction. Cakes are prepared by the *creaming method*, which directly introduces tiny bubbles by aerating the batter, and not by the muffin method. So a cupcake is a small cake rather than a muffin.

Brown writes, 'I have come to the conclusion that the best way (for me) to classify baked goods is by mixing method. Not only does this system make sense, it has made me a better baker' (2004, p. 7). Despite the caveat 'for me', Brown clearly thinks that the taxonomy has a general utility. The advantage he touts most often is that his taxonomy can make one better at baking. 'For instance,' he writes, 'I used to make a really lousy cheesecake until I realized that cheesecake is not a cake, it is a custard pie. Now I treat cheesecake like a custard pie and everything is fine' (2004, p. 7). This advantage is not *merely* practical. As I noted (in Chapter 2 § A), greater manipulative power is often a counterpart to greater understanding of causal structure. In this case, baking prowess results from a better understanding of the underlying substances and processes. So Brown's taxonomy underwrites inductive and explanatory success for accounts of baked goods. Cooking, as Brown practices it, is an applied science – and his categories plausibly satisfy the success clause of the definition of a natural kind.

Of course, cooking is an open-ended creative process. When a chef invents a new dish, the kinds which are appropriate to categorizing the dish might be crafted at the same time that the ingredients and preparation are being devised. As such, there are indefinitely many taxonomies which would successfully account for the dish that is cooked in the end; a possible taxonomy could be made to succeed by commensurate adjustments in the what and how of the cooking. In Chapter 2, I called such categories *fungible kinds*. If the objects that comprise the domain are being crafted at the same time as the analytical story about them, they easily satisfy the success clause. Because of that ease, however, they do not satisfy the restriction clause. So they are not natural kinds.

This rejection is most apt in cases of *fusion cuisine*, where different traditions are combined in novel ways and where there is little constraint from prior practice. Yet Brown is merely attempting to describe baking as he finds it. 'I am not a chef,' he writes, 'I don't have much interest in creating tantalizing new dishes' (2002, p. 1). He suggests that a better title would be 'culinary cartographer' (2002, p. 6); his goal is to chart and understand existing structures of folk gastronomy. Brown writes, 'What I am interested in is making food make sense. I want to understand what makes food tick and how to control the process known as cooking. ... [F]ood is about nothing if not chemistry, physics, math, biology, botany, history, geography, and anthropology' (2002, p. 1). Because Brown is attempting to offer a comprehensive account of baking *as it already exists*, he is not inventing the phenomena at the same time he is describing them. Whatever else it is, **muffin method baked good** is not a fungible kind. Whether or not it does satisfy the restriction clause depends on whether alternative taxonomies would allow for the same success. Since few cooks try to do what Brown does – offer a general explanatory taxonomy – it is hard to say how easy it would be to find alternatives. In order to decide whether Brown has got the right taxonomy or not, we would have to look at the details of how it sorts specific recipes, consider alternatives, and bake more. I suggest, however, that there is at least a *prima facie* case that he has identified natural kinds for baking.

One might object that scones (etc.) are artificial and not natural – so *a fortiori* they cannot form a natural kind. I argued generally against such a move earlier (in Chapter 1 § B.5), but there is a further reply appropriate in this case. The kind **muffin method baked good** is not what Amie Thomasson (2003) calls *essentially artifactual*, because a chef need not have the concept of the kind in order to cook up an instance of it. A chef might cook muffins and waffles with only the usual, nominal way of labeling the products. Contrast Brown's taxonomy, which classifies them together and seems to track real differences between things. The former is merely about how we use words, while the latter reflects the actual causal structure of the domain. We should be able to say something about this important distinction, but we could not if we insisted that a philosophical account of natural kinds cannot apply to anything artificial.

Of course, chemistry itself does not distinguish muffins from biscuits as such. They differ in viscosity and compressibility, perhaps, but general chemistry does not acknowledge **muffin** and **biscuit** as kinds. At the level of subatomic constituents, there is no interesting difference – but neither is there an interesting difference between muffins and mutton at that level. Yet, since natural kinds are specific to domains, the fact that chemistry and physics find no boundary between muffins and biscuits does not show that there is no such boundary in the domain of baked goods and culinary phenomena. This is not to say that baked goods are metaphysical simples or that **muffin** is a primitive universal; remember that the realism I am arguing for is not deep realism, and so I am not making any claims about the fundamental Being that structures the categories. Brown's own account of baking does connect with chemistry; for example, activating gluten by kneading dough is both a phenomenon in the domain of baking and at the level of protein molecules. In short, there is no spooky metaphysics involved in thinking that baking, as a specific domain, might have its own natural kinds.

To conclude, *promiscuity* is not the same as *triviality*. Even though there are indefinitely many natural kinds to be discovered in indefinitely many different domains, this does not amount to radical relativism. A domain like *stuff in pockets* might have no natural kinds whatsoever, because there is no way of categorizing keys and lint which would satisfy both the success and restriction clauses.

C. Agent-relative kinds

Here is a different version of the objection. Posed in an abstract way, it comes to this: There are some domains which include agents who interact with the world, where the interaction involves their dividing the rest of the world into categories. Apart from the activity of these agents, the categories they distinguish lack any unity. Yet an enquiry that attempts to describe the agents might require acknowledging the categories which the agents distinguish. So the agents, just by drawing the distinction, make something a natural kind. This, the objection alleges, is ridiculous.

In § C.1, I describe meerkat signaling behavior. The example is a concrete instance of the abstract formula posed in the objection, with meerkats as the agents in question. The categories distinguished by meerkats themselves do become natural kinds for meerkat ecology – but, I argue, without absurdity resulting. In § C.2, I consider an objection that tries to extend the results from meerkats to imaginary creatures like unicorns. In § C.3, I consider an objection that starts from superstitious humans and alleges that categories like constellations thus become natural kinds. The latter objections fail to appreciate the force of the restriction clause, I argue, and so meerkats are different than unicorns and astrologers.

C.1 Meerkat threats and alarms

Meerkats, also called suricates, are social mammals of the species *Suricata suricatta*. In this section, I draw on the analysis of meerkat alarm calls developed by Marta Manser and others (Manser 2001; Manser, Bell, and Fletcher 2001; Manser, Seyfarth, and Cheney 2002).

Members of a meerkat group take turns standing up as sentinels and sound alarms when threats are spotted. The alarm calls carry both referential information (about the kind of threat) and affective information (about the degree of danger). There are several distinct alarm calls, three of which are of special interest: one is given in response to aerial predators such as eagles and goshawks; a second is given in response to ground predators such as jackals; a third is given in response to snakes such as cobras or adders. A biologist studying meerkats in their natural environment can best understand their behavior by distinguishing the three kinds of signal. So the three calls are (at least plausibly) natural kinds for that domain.

In order to serve in explanations, however, the three calls must be recognized as responses to three different kinds of threat. As Richard Boyd writes of a similar example,

> the referential hypothesis functions as a component in the explanation of an *achievement* – predator avoidance – on the part of the relevant organisms. It helps to explain how the perceptual and cognitive structures in [the organisms] are *accommodated* to relevant causal features of their environment so as to facilitate predator avoidance. (2010, p. 214)

Boyd has in mind ground squirrels which have two distinct kinds of signal, but the point applies as well to meerkats. The causal features of the world to which meerkats respond are differences between kinds of threats in their environment. So, in addition to the natural kind **meerkat aerial alarm** which includes alarm calls of a particular acoustic profile, there is the natural kind **meerkat aerial predator** which includes eagles and goshawks. Understanding meerkats' behavior in their environment requires recognizing both kinds, because the alarm call is explained as a response to and signal of the threat.

I am not suggesting that some meerkat genius intuited the essence of **meerkat aerial predator**, formed a thought, and developed aerial alarm calls to express the thought. The more plausible explanation is that proto-meerkats made some call or other and that natural selection reinforced the connection between the type of sound and the type of threat. There need not have been some preexisting essence in order for there now to be the two kinds. All that needs to be the case is that making sense of situated meerkats *as they are now* requires recognizing alarm calls of a certain kind as a response to threats of a certain kind.

I am also not suggesting that the kind **meerkat aerial predator** has a sharply defined extension. The philosophical imagination can easily invent boundary cases: e.g. imagine a robot that looks like an eagle but which, if it actually catches a meerkat, rewards the meerkat with food and a gentle massage. Here I reiterate a lesson from Chapter 1, that natural kinds need not have sharp or timeless boundaries.

Manser (2001) labels the first two kinds of calls 'aerial' and 'terrestrial' alarm calls, but the third is not the 'snake' call. The third call is given not only in response to snakes but also to physical signs of predators (such as stool or urine) and to meerkats from other groups. If we think of meerkats as themselves biologists, devising a three-word vocabulary for some higher level biological taxa in their environment, then it seems as if they have made a horrible mess in the third case. Yet meerkats are not little biologists. The three calls function both referentially and motivationally; alarm calls are responses *to* something in the environment and imperatives *for* other meerkats to act. When threatened by aerial predators or large terrestrial predators, such as jackals, meerkats flee down bolt holes. In response to a snake, however, they can group together to scare off or overcome the predator. A similar response is appropriate when jackal droppings are sighted; other meerkats should come over and look around. And the same response is useful when rival meerkat groups are nearby. So Manser calls this third kind of call a 'recruitment' alarm call.

I argued above that **meerkat aerial alarm** and **meerkat aerial predator** are natural kinds for enquiry concerning meerkats in their natural environment. Because recruitment calls are used in *prima facie* disparate ways, it may be tempting to resist a parallel argument for the conclusion that **meerkat recruitment alarm** and **meerkat recruitment threat** are natural kinds. An ornithologist may not have any word for just the category **meerkat aerial predator**, but at least it includes only birds. The category **meerkat recruitment threat** includes snakes of several varieties but also (among other things) jackal poop. The temptation is to think that there could not be any natural kind that includes such disparate stuff. Part of this temptation comes from thinking of the domain of herpetology when thinking about a kind that includes snakes. Of course, if the domain of enquiry just includes snakes and jackals, then there is no natural kind common to both adders and jackal poop. However, it would be a mistake to think that ornithology or herpetology is the domain at issue here. The meerkat alarm call kinds would not be natural kinds for a domain that did not include meerkats, so the referential counterparts in the meerkats' environment would not be either; the taxonomy of threats only serves prediction and explanation along with the taxonomy of alarm calls, and even then only for a domain which includes meerkats and their natural environment. It may be surprising that the domain has a natural kind which includes both snakes and jackal poop, but it has these natural kinds because of the specific relations between meerkats and those things.

A corollary of this is that the threat-kinds would not be natural kinds if meerkats had never existed. Paul Churchland argues that this contingency means that the kinds are in some way arbitrary. He writes, 'One does not naturally think of biological species as being as arbitrary as I am here insisting, since our world presents us with only a fixed subset of the infinite number of possible species. But that subset is an accident of evolutionary history.' So, Churchland concludes, 'tiger, elm, and apple' – we can add meerkats – 'turn out to be merely practical kinds' (1985*a*, p. 12). If the course of evolution had gone differently, such that there were no meerkats, different meerkat predators, or even just a different meerkat signaling system, then neither the alarm calls nor threat classes that Manser et al. identify would be features of the world. Yet, as I argued in Chapter 4 § A, Churchland's dilemma of real versus practical kinds presumes a false dichotomy. Given meerkats as they have actually evolved, there are natural kinds for domains of enquiry that include them. These natural kinds reflect the real structure of meerkat ecology.

C.2 Unicorns and fictobiology

In the previous section, I argued both that meerkat alarm calls and the kinds of things which prompt meerkats to sound the calls count as natural kinds. As a corollary, I suggested that those would not be natural kinds if meerkats did not exist. In this section, I consider an objection which challenges the corollary. Meerkat alarm calls and threats would not be natural kinds for a domain of *actual organisms* if meerkats did not exist, but we might enquire into a counterfactual domain of *possible organisms*. The objection alleges that the alarm calls and threats would be natural kinds for such a domain *regardless* of what the world is like. So, it concludes, my account recognizes unreal kinds as natural.

Before answering the objection, I want to get clear on just what is being alleged. If meerkats were extinct, domains of present ecology or biology would not need meerkat kinds. One might try to do historical ecology and uncover structures relevant to extinct species. If this were *paleobiology* in a strict sense, concerned with fossil meerkats, then meerkat signaling behavior would not be among the phenomena in the domain of enquiry. Fossils do not preserve stimulus–response relationships or the acoustic profiles of signals. There is nothing that prohibits a broader enquiry into extinct species, of course. We can coherently ask whether or not there are any natural kinds particular to the behavior of *Tyrannosaurus rex*. Such an enquiry simply suffers from a dearth of evidence, and so we will probably never be able to answer such questions. On my account, there will be natural kinds in domains even if we have limited or no empirical access. This *requires* that natural kinds be real features of the world. There are no existent members of the species *T. rex*, but the natural kinds for *T. rex* behavior are determined by how such members did behave (in the past) or would behave (if they existed now). So merely considering extinct organisms will not motivate worries about triviality.

One might instead ask after natural kinds for organisms which never existed but (in some sense) could have existed. Let's take unicorns as our example. We can ask how unicorns *would* behave if they *did* exist. The question is coherent. This would not be paleobiology, but instead what we might call *fictobiology* – the biology of fictional organisms. I do not put any constraint on what can count as a domain of enquiry, so we can ask whether there are natural kinds for fictobiology. The objection is that **unicorn** and **unicorn food** would be natural kinds for fictobiology and that it is absurd to say that those are natural kinds.

I argue, contrary to the objection, that my account does not recognize any natural kinds for fictobiology. Start with **unicorn**. Since there are no unicorns in the world and never have been, the kind does not apply to anything. An element that has never been produced may be a natural kind for physics and chemistry, based on the systematic nature of the periodic table. The element of atomic number 110, for example, could figure in prediction and explanation even before it had been produced; so **Darmstadtium** was a natural kind even when it did not apply to anything. Similarly, a compound that has never been synthesized might be a natural kind for chemistry. So the point is not that a kind must have actual members in order to be a natural kind. Rather, an empty kind must have systematic connections to what does exist in the domain in order to count as a natural kind. Unicorns do not have any systematic connection to actual organisms. So unicorns would not be a natural kind.

Even though there are no unicorns, there are actual things in the world that could be unicorn food: oats and hay, for example. This does not figure in the explanation of any actual phenomena. Rather, it only explains *possible* phenomena: that a unicorn would be happy with a bucket of oats, for example. There are at best weak explanations in terms of the category **unicorn food**. Moreover, the possible phenomena can only be made determinate by stipulation. After all, there is only the analogy with horses that suggests that unicorns would eat oats. A quick search of the internet turns up other speculations: that unicorns generate solar power by means of their horns; that they eat human flesh and the blood of children; that they subsist on stones, tree sap, and rainbow glitter; and so on. Since unicorns are imaginary creatures, we can imagine them to eat anything whatsoever. So the kind **unicorn food** will figure in its weakly successful explanations just in virtue of our specification of how we are going imagine unicorns. It is thus, at most, a fungible kind. The specification of fictobiology is too fluid for the enquiry to have any natural kinds whatsoever.

C.3 Constellations

The enquiry of fictobiology fails to have natural kinds because the organisms it addresses are utterly imaginary. A final version of the objection starts with real organisms and considers kinds which those organisms imagine. Such a case would develop if all of the birds were removed from the meerkats' environment, if they continued to sound the aerial alarm call and scramble into bolt holes as a matter of playacting. Rather than consider a thought experiment like that, however, we can ask about superstitious humans who identify constellations. The domain in question is anthropology of a past or present group of people who recognize specific patterns among the stars.

Constellations are perhaps the canonical example of a kind of thing which must *not* count as a natural kind. Sam Page provides the standard rationale:

> We, or more specifically our ancestors, determined which stars comprise which constellations. We can come up with new constellations whenever we like simply by pointing out a few stars and giving the cluster a name. ... Though it is *prima facie* plausible that reality is individuated intrinsically into stars, reality is not individuated intrinsically into constellations, since it is people who divide the night sky into constellations. (2006, p. 328)

The central idea is that we can make up constellations however we like. Imagine, for example, that I pick a cluster of visible stars and specify them to be the new constellation Gedankion. I might then do some research to learn which of the stars in Gedankion is furthest from the Earth, which is brightest, and so on. I might make a regular habit of going out at night, pointing up at the sky, and saying things like 'There it is!' Notice that the kind **Gedankion star** will not figure in any explanations of facts about those stars. So it will not be a natural kind for astronomy. If someone asks me what I am doing, when I am pointing at the sky, I might explain my behavior in terms of that kind: 'I am pointing at that star, rather than some other star, because it is a star in Gedankion.' So the enquiry in which the kind figures would include me, as a signaling organism. Since the stars were just picked arbitrarily, however, the category **Gedankion star** is at best a fungible kind. If I had picked different stars, then I would point at different ones later. Success is assured by the tight connection between the specification of the kind and the explanations that employ it.

One might argue that traditional constellations are not so loose. There was a point at which stars in the Big Dipper were picked more or less arbitrarily, but now there is a tradition. If someone goes out and points to the Big Dipper, they are not themselves stipulating the kind. So, one might argue, the Big Dipper is not a fungible kind in the same way that the novel and arbitrary constellation Gedankion would be. Robert Schwartz uses the example of constellations in arguing that we make kinds, as part of an argument against 'the invidious metaphysical distinction between artificial and genuine kinds' (2000, p. 156). There are truths about the Big Dipper which do convey 'objective information about the skies'; for example, the star Zubenelgenubi is not part of the Big Dipper. Schwartz insists, 'Neither conventionality nor failure to occur in the basic laws of physics or astronomy seem to preclude the properties **constellation** and **Big Dipper star** from having a real purchase on "reality"' (2000, pp. 155–6, my bold).

It is unclear what predictive or explanatory work the kinds **constellation** and **Big Dipper star** are meant to do. There are several ways to construe the example.

First, there might be a brute tendency for people to identify constellations. As such, some constellations or other would figure in anthropological explanations. Different cultures identify different constellations, but the general category would figure in a cross-cultural regularity. As such, **constellation** would be a natural kind for anthropology. This would not extend to specific constellation-kinds, like **Big Dipper star**, unless those constellations were recognized universally.

Second, constellations can be used as aids in navigation and as guides to the seasons. Schwartz writes, 'In navigating the land and seas it is helpful, for example, to be able to locate the Big Dipper and know the outermost star of its bowl points toward the North

Star' (2000, p. 155). Here the anthropological situation is interesting. Edwin Hutchins describes the local 'magician cum astronomer' on Boyowa Island in the Trobriand Islands of Papua New Guinea. The traditional practice allows the magician-astronomer to determine the agricultural calendar based on the visibility of specific stars in the sky just before dawn. Making this determination requires attention to 'a large number of named constellations'; constellations are more useful for this than a single star, because is possible to recognize constellations even if some of the stars in it are obscured by clouds (2008, p. 2012). Hutchins also gives examples from Micronesian navigation (1995, ch. 2; 2008). This provides further reason to think that **constellation** as a general category corresponds to important features of the anthropological domain.

Hutchins suggests that there may be some cross-cultural regularities as to *which* constellations are identified. He explains that 'some of the constellations recognized in the Trobriand system roughly match constellations identified in our own tradition. Some of these similarities suggest the operation of the gestalt laws of continuity and proximity.' The similarities that there are in which constellations are identified thus result from general cognitive patterns, and so need not involve *sui generis* kinds. Moreover, different groups do not identify precisely the same constellations; Hutchins adds that, 'cross-cultural variability in the composition of constellations shows that gestalt laws are not sufficient to account for the observed culturally specific groupings' (2008, p. 2018, fn. 2). The strongest anthropological regularity is just that people will identify *some* constellations. There are weaker regularities in which kinds they identify, and those are explained in terms of general cognitive tendencies. So again it still looks as if **constellation** might be a natural kind but **Big Dipper star** will not be.

Third, contemporary astronomers talk about constellations. Schwartz offers this as further evidence for the reality of constellations; he notes that 'current astronomy divides the sky into regions associated with prominent constellations. These constellation based boundaries provide a standard scientific map for locating objects in the heavens' (2000, p. 155). Examples of such usage are not hard to find. A NASA webpage, for example, describes Zubenelgenubi as 'the second brightest star in the constellation Libra' (NASA 2004). Although the Big Dipper and Libra do figure in successful astronomical practice in this way, astronomers would do just as well with an alternative system of constellations. The system they use has only the advantage of familiarity.

To return to a previous example (from the end of Chapter 2), brands of instruments are sometimes mentioned by name in scientific reports. Since other brands might have been used instead, the kinds fail to satisfy the restriction clause. They are fungible rather than natural kinds. The case is parallel for constellations being mentioned by name, where they are serving simply as instruments for referring to a specific swath of sky. They do not thereby become natural kinds.

Fourth, astrologers talk about the influence that constellations and planets have on Earthly life. To his credit, Schwartz does not try to defend this as a real influence. Jupiter has no influence on my love life, and so **the influence of Jupiter on my love life** is not a natural kind. As I noted above, empty categories can be natural kinds if they occupy an important systematic place in a generally successful taxonomy. The null influence of Jupiter does not do so for astrology, because astrology has not got a generally successful taxonomy. One might, I suppose, shift attention to the enquiry of *debunking astrology*. For that largely negative endeavor it may be important that there is no such influence. Acknowledging that the absence of such influence is a real feature of the world is no embarrassment, however. If there are kinds that must be recognized in order to debunk astrology, then friends of science should welcome an account of natural kinds that recognizes them.

In sum, my account of natural kinds is not forced to recognize specific constellations as things *out there* in the sky.

D. Coda on promiscuity

In this chapter, I have argued that my account of natural kinds is not too promiscuous. In the course of doing so, I have been open to kinds like **muffin method baked good** (for the domain of baking), **meerkat recruitment threat** (for meerkat ecology), and **constellation** (in the general sense, for anthropology). Whether these are genuinely natural kinds depends on the facts of those domains. In baking, for example, I am not sure whether the kinds do better than possible rivals and so satisfy the restriction clause. The fact that this is an open, empirical question makes clear the sense in which natural kinds depend on the actual structure of the domains in question.

Some accounts of natural kinds are more chaste and only acknowledge the sparse ontology offered by our most fundamental physics. If we were to restrict 'natural kind' in such a way, then we would still need a term for the real structures of domains which I have been describing.

6
Causal Processes and Property Clusters

This chapter is not initially about the account of natural kinds which I have developed and defended, but instead about homeostatic property clusters. Roughly, a homeostatic property cluster (HPC) is a stable collection of properties which tend to occur together, unified by the causal tendency for them to occur together. Richard Samuels and Michael Ferreira write that 'philosophers of science have, in recent years, reached a consensus – or as close to consensus as philosophers ever get – according to which natural kinds are *Homeostatic Property Clusters*' (2010, p. 222).

If we take this as a definition, 'natural kind = HPC', then it could have been listed among the common assumptions about natural kinds back in Chapter 1. The definition is easily dismissed, however, because there clearly are natural kinds which are not HPCs; e.g. **electron** is a natural kind even though electrons, as fundamental particles, have the basic properties they do without any underlying causal process to hold them together.

Moreover, HPCs were not originally introduced as a conception of natural kinds. Rather, they were introduced as an explanation of what some specific kinds actually are in the world. Those natural kinds turn out to be HPCs. This is the value of discussing HPCs after we have the account of natural kinds in place: Many natural kinds are HPCs, and we can understand them better by recognizing this.

Much of this chapter is concerned with examples of species in biology. I have already argued (in Chapter 3) that many species taxa are natural kinds. Here I argue that they are HPCs and, moreover, that recognizing this allows us to understand more about them as natural kinds. Exploring the example in some detail requires considering some points which are just about species, but it also raises two important points about HPCs in general: First, the properties which are held together by the HPC need not just be ones which are typical of every member of the kind. Even if we initially notice the kind on the basis of superficial similarity between some of its members, the process which maintains that similarity may also implicate other individuals which are superficially very different. So the structure of properties explained by the HPC includes similarities between members but also the systematic differences among them. Second, there is a difference between HPCs which are maintained by a single causal process and those that are maintained by similar but independent causal processes. I call the former *token-HPCs* and the latter *type-HPCs*. Members of a biological species, connected by shared history, form a token-HPC. Chemical kinds, like water, form a type-HPC if they form an HPC at all.

In § A, I take up an objection that HPCs are ill-fitted to account for polymorphic species. Advocates of the HPC approach have answered the objection by saying that all members of a polymorphic species do have things in common, namely dispositions or conditional properties. I argue that this response fails to deflect the objection and also fits poorly with the HPC account itself. An HPC is better understood as unified both by the underlying causal mechanism that maintains it and the resulting, potentially disparate constellation of properties, so instances of an HPC kind need not all have similar intrinsic properties. The causal mechanism can produce and explain both similarities and systematic differences between the kind's members.

In § B, I reflect on species to motivate the distinction between token-HPCs and type-HPCs.

In § § C–D, I consider the relationship between species as HPCs and various aspects of the species problem. I first argue that the account of species as HPCs reinforces conclusions about the species problem that I argued for in Chapter 3. I then consider *individualism* about species – a metaphysical view which most take to be a rival to the HPC account. Once we recognize the distinction between token-HPCs and type-HPCs, it becomes clear that a historical individual just is a token-HPC.

In § E, I turn to examples beyond species. Water, which is often bandied about as something which has unproblematically got an essence, is plausibly better seen as a type-HPC. Yet shifting from token-HPCs (like species) to type-HPCs introduces new problems. I do not have a fully developed account of type-HPCs, and I end by pointing to where difficulties remain.

A. Species as the specimen of an HPC

Richard Boyd (1988) introduced *homeostatic property clusters* (HPCs) in an article about moral realism, where he offered it as way to make sense of philosophical kinds like *moral goodness*. He also meant it to account for *rationality* and *reference* as features of scientific practice, and the definition of HPCs given in the context of scientific realism echoes the one given in the discussion of moral realism (1989). Nevertheless, biological species have been the clearest examples of HPCs.

Note that Boyd never *defines* natural kinds as HPCs. Rather, as I discussed in Chapter 4 § A.1, he gives an account of natural kinds as categories that scientists use to accommodate the world. Because the world has a causal structure, many natural kinds end up being HPCs. As we will see, the same is true on my account of natural kinds. Although I differ with Boyd on some important points, I agree that species taxa in biology are best understood as HPCs. However, we should recognize that species are in some ways the easy case. Under a suitably wide lens, every member of a species participates not just in similar causal processes but in the very same causal process – every single member of the species is a product of its shared evolutionary history. Other natural kinds do not have the same unity and so cannot as easily be understood as HPCs.

So let's begin with biological species. As an example, consider mallards – birds of the species *Anas platyrhynchos*. Members of the species typically look like ducks, walk like ducks, and quack like ducks. It is an aphorism that a thing which possesses these properties is probably a duck. We can extend the list to include more precise properties, such as physiological and biochemical ones, but for the sake of illustration I will continue to talk in terms of gross morphology. Regardless of how thorough we make our list, we will never arrive at a list of occurrent properties such that possession of them is both necessary and sufficient for being a duck. A one-legged mallard will not walk like a duck, and a mute mallard will not quack – but both are nonetheless still mallards. So the properties are not essential to the species. Rather, the properties form a cluster.

It is not merely a convenience to think of the properties which are typical of ducks as a cluster. An individual duck is a locus of properties not by chance, but rather because of facts about her physiology. The properties typical of ducks occur in her because of a causal process. The properties obtain – and continue to obtain into her dotage – literally because of her homeostatic properties. Her present physiological condition was brought about by the related causal process of her growth and development, from egg to duckling to full-grown duck. Here the 'homeostasis' is less literal, because development involves change rather than mere maintenance. (Paul Griffiths (1999) emphasizes the importance of development within species conceived as HPCs.)

Yet our concern is not just with an individual duck. There are, indeed, many individual ducks in the world. These separate loci of properties from the cluster are each held together by their own (similar but numerically distinct) physiologies. The numerous ducks are also the result of a causal process. The point is not simply that there is a similar but distinct process going on in each duck, but rather that ducks altogether are the result of a single, shared natural history. To put it simply, the population of mommy ducks and daddy ducks spawns clusters of duck properties in the form of ducklings. Those mommy and daddy ducks themselves were spawned by a previous generation, and so on, back through the whole evolutionary lineage.

One might pedantically object that there is no such thing as a *daddy duck*. Males of the species are *drakes* or simply *mallards*. Indeed, the word 'mallard' originally meant a wild drake, and some people I have spoken with still use the word that way. My sense of contemporary English is that both 'mallard' and 'duck' can be used in sex-neutral ways. This shift would be important if the question of natural kinds was primarily a matter of understanding natural kind *terms* – if, in the Kripke–Putnam tradition, one hoped to employ reference to natural kinds in order to solve semantic or metaphysical problems. Even Boyd sometimes presents the question as one of language, writing 'that the subject matter of the theory of natural kinds is ... the use of natural kind terms and concepts' (Boyd 1999*a*, p. 148). Yet he admits that the use of a particular term to mark a particular category is often a matter of convenience. When we might adopt one terminology or another, what matters is that 'either choice would result in the establishment of a vocabulary ... in which the same class of causally and explanatorily relevant distinctions could be drawn' (Boyd 1999a, p. 158). The taxonomic system is what matters, not the mapping of linguistic labels onto pigeonholes in the taxonomy. In what follows, I will use 'mallards' as the common name for the species, 'duck' as the word for a female of the species, and 'drake' as the word for a male. These may be read as *'Anas platyrhynchos'*, *'A. platyrhynchos* female', and *'A. platyrhynchos* male' respectively.

Although insistence on the difference between the words 'ducks' and 'drakes' is a pedantic point, insistence on the difference between the creatures ducks and drakes is not. Mallards importantly come in two varieties, differing according to sex. An HPC identifies the causal process that explains the *similarity* among members of the species, but what does it have to say about *differences*? This question motivates an objection to the view that species are HPCs. I turn to that next.

A.1 Worries about polymorphism

Marc Ereshefsky and Mohan Matthen (2005, henceforth E&M) raise an objection to the HPC approach. Although I do not think that the objection ultimately succeeds, I think that the usual responses to it are inadequate.

It is uncontroversial that even judgments of surface similarity require a prior structuring on qualities. The structure might be common sense perception or a sophisticated theory, but it determines the relevant properties and their respective importance. Things count as similar or different relative to that determination.

Here E&M are just presenting a standard image of natural kinds. For example, C. D. Broad (1920) imagines individuals as a fluid filling the space of properties and suggests that natural kinds correspond to dense lumps, where many individuals share the same properties. Quine (1969a) emphasizes that natural kinds presume a space of similarity and says that a kind corresponds to a 'qualitatively spherical' region. As E&M put the point: Given a metric on the space of qualities, similarity can be defined as proximity according to that metric. They call this space of qualities with its accompanying metric a *morphospace*.

As they understand the HPC approach, biological taxa are identified in three steps. (Although I paraphrase each step, they enumerate the steps in this way; p. 5.)

1. We look around and find creatures that are clumped together in morphospace. The properties of these creatures form clusters.
2. We look for the mechanisms responsible for these clusters.
3. We rejigger the morphospace in light of these mechanisms and start again.

The final step is necessary, they say, because we ultimately want to group together specimens that do not initially look similar. For example, imagine we start by making brief observations of caterpillars and moths. The two occupy different regions of the initial morphospace; i.e. they do not count as being *like* one another. Yet we want to count *Biston betularia* caterpillars and peppered moths as members of the same species, because the caterpillars grow up to be peppered moths who lay eggs for such caterpillars. Trying to understand this in terms of property clusters and similarity is awkward. The twig-like, crawling caterpillar and the speckled, flying moth are different in their surface features. If we want to say that they form a natural kind – the argument goes – then there must be many properties that they hold in common. That is, we need to unify the list of properties which are clustered together in the HPC.

There are several possible options, none of them especially appealing.

First, one might refuse to rejigger the morphospace at all. This would require saying that the caterpillar and the moth are members of distinct HPCs. E&M call this the *splitting maneuver* (p. 12). For the account that species are HPCs, it would require saying that the caterpillar and moth are not members of the same species – a biological absurdity.

Second, one might rejigger the weighting of properties in the morphospace. Suppose we do this with moths and say that flying is the more significant property, and that the natural condition for a peppered moth is to be a flying creature. Any individual of the species which cannot fly would count as a monster, in the sense of being deformed. We might call this the *standardize maneuver*. This would be wrongheaded, because a caterpillar's not flying is different than a one-legged duck's not walking. The crawling caterpillar is doing just what a member of its species naturally does *at that point in its life cycle*. Moreover, even if it were acceptable to say that a caterpillar is a deficient moth, such a maneuver is clearly wrongheaded for handling sexual dimorphism. Neither ducks nor drakes are deformed counterparts of the other. Rather, both instantiate perfectly legitimate ways to be mallards. Obviously parallel points can be made about human sexuality – neither women nor men are more fully human than the other.

Third, one might rejigger the morphospace by changing the properties that are included in it. The caterpillars and moths that are members of the species *B. betularia* neither all crawl nor do they all fly, so one might instead say that they have the disjunctive property crawling-or-flying. E&M call this the *mereological maneuver*. As they note, it threatens to make similarity vacuous. They write:

> One needs ... to be careful how one uses such powerful logical tools: they can become a universal solvent that makes all variation disappear and collapses the entire biological domain into a single morphoclump. For one could account for the differences between bees and apes by a function on some suitably selected variables such as genetic constitution and environmental inputs ... And one could then regard the vast network of ecological relationships that constitute the biosphere as a single homeostatic property cluster maintaining polymorphism in 'Gaia'. (p. 9)

The point is that any two populations share stitched-together disjunctive properties. Moreover, the approach loses track of the fact that there are two separate clusters. Members of *B. betularia* are not just crawling-or-flying and cocoon-weaving-or-egg-laying. The crawling goes together with the cocoon weaving. An HPC *qua* cluster of properties must somehow encode the fact that members of the kind come in distinct, coherent varieties.

Despite the difficulties with this approach, advocates of the HPC view of species have typically opted for something like the mereological maneuver. Writing before E&M, Boyd says:

> The fact that there is substantial sexual dimorphism in many species and the fact that there are often profound differences between

> the phenotypic properties of members of the same species at different stages of their life histories (for example, in insect species), together require that we characterize the [HPC] associated with a biological species as containing lots of conditionally specified dispositional properties for which canonical descriptions might be something like, 'if male and in the first molt, P', or 'if female and in the aquatic stage, Q'. (1999a, p. 165)

Note that Boyd does not rejigger the morphospace to introduce disjunctive properties (like crawling-or-flying) but rather to introduce conditional properties (like crawling-if-larval). The maneuver is not so much *mereological* as *dispositional*. Robert Wilson, Matthew Barker, and Ingo Brigandt also give the dispositional reply in response to E&M, concluding that there is 'nothing in what [E&M] say about this that makes this a problematic, implausible, or ad hoc view to adopt. Indeed, biologists' knowledge about species reflects the empirical presence of complex and conditional traits' (2007, p. 211).

There may be some sense to saying, of an individual caterpillar, that it has the dispositional property *if adult, then flying*. That individual caterpillar will grow up to be a moth, after all. Regardless, the dispositional maneuver is hopeless in typical cases of sexual dimorphism. Consider again a duck. She will never be a male, so the property *if male, then green-headed* could only be an unrealized disposition. It depends on the counterfactual that *if* she were a male, *then* she would have a green head. How might this counterfactual be understood such that it would be true of her? I can think of three ways.

First: *Male mallards have green heads. So, if she were a male mallard, then she would have a green head.* This counterfactual is true, but will be of no help to the HPC view. It is a syllogism. It is true of the telephone on my desk as much as of the duck that if *it* were a drake then it would have a green head.

Second: *She is a mallard. Male mallards have green heads. So, if she were a male, then she would have a green head.* The idea is to hold her species fixed when evaluating the counterfactual. The counterfactual is true, I guess, but only because we have an independent grip on what counts as a mallard. So it would be circular to say that properties like this are constitutive of the HPC that *constitutes* the species. The species boundaries must already be in place before this counterfactual has a determinate truth value. Furthermore, I do not see how the counterfactual corresponds to a dispositional trait of the duck. What makes it true is not any trait of the duck herself but facts about other mallards – the male ones.

Third: *If we intervened on her sex so as to make her a male mallard, then she would have a green head.* This might be true, but I am unconvinced. There are many ways we might perform a sex change operation on a duck, some of which would influence her plumage and some of which would not. Perhaps the regimen of hormone treatments would in fact turn her head feathers green, but I do not know enough about ducks to say. It is possible that no ornithologist does. This ignorance does not make us question whether ducks and drakes are all mallards, but it does undo any hope of cashing out a dispositional property in terms of an intervention counterfactual.

Admittedly, it is possible to make sense of counterfactuals like these for some organisms. The bluehead wrasse (*Thalassoma bifasciatum*), for example, is sequentially hermaphroditic. A female bluehead may later change into a male. Just as we can say of an individual caterpillar that it will develop into a moth, we can say of a female bluehead that it could become a male. And just as we can describe the properties the eventual moth will have, we can describe properties that the secondary male fish would have: The female-turned-male bluehead wrasse would have a blue head. So it makes sense to say, of the female, that it has the *disposition* to be blueheaded on the condition of being male. This works because there is a typical way in which females of the species become males, so the question of what this individual would be like if male has a well-defined answer.

The case of the duck is less clear because there is no typical way for a duck – once it is a duck – to become a drake. Even if there were a way of resolving the counterfactual, however, it is not clear what justifies the metaphysical backflips. The conditional properties are only introduced to fulfill an apparent bookkeeping requirement of the HPC approach, namely that there be a unitary cluster of typical properties.

Taking a step back from the specific objection, it is unclear why conditional properties like green-head-if-male should have been part of the HPC approach at all. Consider the cluster of properties that are typical of all mallards: two legs, two wings, a beak, quacking behavior, and so on. The initial motivation for treating this as a property cluster was that these features, although typical, were not necessary. A one-legged duck which lost its leg when attacked by a fox is still unproblematically a duck, even though it lacks the property of having two legs. The dispositional maneuver would allow us to say that even the one-legged duck has the property *two-legged-unless-it-loses-a-leg*. It is true of the duck that it would still have two legs if it had not been attacked, so the corresponding counterfactual is true. Yet the *disposition* to have two legs becomes a *necessary* feature of mallards, a disposition that is just unrealized or defeated in one-legged ducks. Although the HPC approach began with the idea that species could be distinguished (in part) by properties without having the properties be necessary conditions for species membership, the dispositional maneuver cooks up conditional properties which are necessary for membership.

I think that this is a real tension in expressions of the HPC account. Where an HPC does not involve instances all alike, Brigandt suggests that it 'maintains ... a *characteristic distribution and correlation* of ... features' (2009, p. 83). Yet he assumes without argument that *something* must be shared by members of the kind, and he explains what is shared by means of the dispositional maneuver: 'What is shared are complex and conditional properties such as "if female then B" or "if larva then C"' (Brigandt 2009, p. 83).

So I think E&M's objection is successful against the characterizations of species as HPCs that have been offered. In the next section, I suggest how.

A.2 Getting over similarity fetishism

Natural kinds have traditionally been conceived as collections of things that share many features. Because all the members of the kind are similar, one can make successful inductions from one or a few members of the kind to all members of the kind. This basic idea, that a natural kind is primarily a set of similar things, is *similarity fetishism*. (We've seen this before; Chapter 1 § B.1.)

Boyd begins with worries about projectibility. Projectible predicates are those that allow for successful inductive inferences; they correspond to properties that can be observed in one individual which are typical of all individuals. So the task of charting a natural kind becomes a matter of writing down the list of projectible predicates. There is one list, and it becomes the property cluster of the corresponding HPC kind. As a result, Boyd writes, 'Species are defined, according to the HPC conception, by those shared properties and by the mechanisms (including both "external" mechanisms and genetic transmission) which sustain their homeostasis' (1999c, p. 81). He says also, 'I do not, for better or worse, hold that HPC kinds are defined by reference to historical relations among the members, rather than by reference to their shared properties' (1999c, p. 80). Ereshefsky calls him to task for this, quoting this same passage and concluding, 'For Boyd, similarity trumps historical connectedness' (2007, p. 296).

Crucially, similarity fetishism and the impulse behind the HPC view are distinct. If we deny that the identification of HPCs is a matter of drawing up *one* list of projectible predicates for the kind, then there need not be the recursive rejiggering of the morphospace that E&M describe. Of course, E&M are right to say that there is no utterly objective, non-trivial sense of *similarity*. We begin with some characterization of the space of qualities, and we identify property clusters. We then explain those clusters in terms of causal mechanisms. Perhaps we will be able to characterize the kind that is sustained by that mechanism in terms of a single list of canonical properties, but perhaps not. When we look at how the properties of an adult duck come to be manifest in a particular individual, we recognize that she had to be a duckling first. There is a related but not entirely coincident cluster of properties that characterize ducklings.

Similarly, when we consider how there come to be ducks in the world, we recognize that the process both *requires ducks and drakes* as input and *produces ducks and drakes* as output. Again, drakes are characterized by a cluster of properties that is related to but not entirely coincident with the one that characterizes ducks. The story of mallards is one that involves ducks, drakes, ducklings, and baby drakes (drakelings?).

The difference between the process I described and E&M's description of how HPCs are identified is at the last step. My version might be summarized in this way:

1. We look around and find clusters of properties.
2. We look for the mechanisms responsible for these clusters.
3. We identify natural kinds by the scope of those mechanisms.

Note that 3 is not the end of enquiry. We can still study the kind that we have identified. We may even have to modify our understanding of the kind in light of further discovery. The point is simply that we are not required to immediately reconstrue the properties involved so that all the members of the kind are clustered together in morphospace; i.e. so that there is a single list of properties which are all typical for members of the kind.

I have acknowledged that ducks and drakes are distinct in important ways, so one may object: Since stronger inductive conclusions can be drawn about *ducks* by examining a duck then can be drawn about *mallards*, what I have said makes **duck** and **drake** separate natural kinds. The objection is that I have performed what E&M call the splitting maneuver while propping the two kinds close together in an effort to hide the seam. Although the objection would make sense if we were trying to determine whether ducks and drakes are the same kind of thing *tout court*, there is no reason to suppose that the duck must only be counted as a member of one natural kind.

Moreover, the separate kinds **duck** and **drake** both rely on the general causal mechanism of the species, reproducing generation after generation. The homeostatic mechanism which keeps the world full of duck properties clustered together in the form of ducks is the same one that propagates drakes. So the two separate kinds, considered as HPCs, both require the same mechanism. The mechanism is not just the physiological process of an individual – although it includes that – but the whole reproduction and selection history of the species. If we begin by noticing one or the other cluster of properties, we are led to acknowledge a causal story that explains both.

E&M resist an attempt to appeal to a single mechanism, on the grounds that there will not be a single developmental mechanism present both in males and females, such that it makes males exhibit male features and females exhibit female features. They write that 'the HPC theorist would need to find an underlying similarity between males and females that expresses itself differently in different circumstances. But there is no such similarity in mammalian [or, we might add, avian] species' (p. 9). This works as an objection to the dispositional maneuver, which attempts to explain dimorphism by inventing properties that are typically present in both the ducks and the drakes. Even though there is no single causal process present in a duck which is responsible for its duckness and also present in a drake which is responsible for its drakeness, the duck and drake do participate together in a *larger* process. The genetics and development of ducks and drakes differ, but they are both part of the mallard reproductive process by which a new generation of mallards (both ducks and drakes) is produced. And all mallards are products of the causal selection history which produced the species.

So I am suggesting that we understand species as an HPC kind not as Boyd does but instead 'by reference to historical relations among the members' (Boyd 1999c, p. 80, quoted above). The suggestion that species boundaries are produced by causal processes that might be called 'homeostatic' is not unique to Boyd, after all. Niles Eldredge and Stephen Jay Gould ask what marks the limits of species – that is, why some separated populations become separate species but others do not – and they give their answer in terms of 'homeostasis'. They write:

> The answer probably lies in a view of species and individuals as homeostatic systems – as amazingly well-buffered to resist change and maintain stability in the face of disturbing influences. ... The coherence of species, therefore, is not maintained by interaction among its members (gene flow). It emerges, rather, as an historical consequence of the species' origin as a peripherally isolated population that acquired its own powerful *homeostatic* system. (1972, p. 114, my emphasis)

Figure 6.1 A monstrous denizen of the deep sea; a female anglerfish of the species *Linophryne lucifer* (Goode and Beane 1895, plate 121).

The causal process that unifies the species operates on the population. The homeostasis is not the production of approximately identical individuals but the maintenance of a stable configuration among individuals. The continuation of mallards, as a species, requires comparable numbers of ducks and drakes.

One might agree with this construal of HPC species but insist that it just severs the connection between HPCs and natural kinds. In the next section, I will articulate and then try to answer this objection.

A.3 Natural kinds and systematic explanation

By considering mallards and peppered moths, I have perhaps been helping myself to the easy cases. Consider instead the extreme dimorphism of some species of deep-sea anglerfish. The females considered separately are rather exotic, with bioluminescent filaments growing out of their heads. Looking at an image like Figure 6.1, it is easy to understand how they came to be called *seadevils*. However, the males are much smaller and less fearsome. They are incapable of feeding and attach to the females for reproduction. In some species, this attachment is permanent. As Theodore Pietsch and Christopher Kenaley explain, 'the male's attachment to the female is followed by fusion of epidermal and dermal tissues and, eventually, by a connection of the circulatory systems so that the male becomes permanently dependent on the female for blood-transported nutrients, while the host female becomes a kind of self-fertilizing hermaphrodite' (2007) (see also Pietsch 2005, 2009).

The difference is much more radical than that between ducks and drakes. Take a specific seadevil species, such as *Linophryne arborifera* (see Figure 6.2). Females and males are so dissimilar that there are few inductions one can make about the species in general from a single sample. If one were simply looking for projectible predicates, then the species would not be a relevant kind at all. Rather, one might instead construct the kind *female anglerfish*. Although a diverse lot, one can draw more inductive conclusions about them from a single specimen than one can about members of a specific anglerfish species.

If we are in the grip of similarity fetishism, this situation should be quite disturbing. *L. arborifera* females share a distinct, recognizable cluster of properties. They can be distinguished from the females of other anglerfish species. There is also a cluster of properties associated with anglerfish of both sexes. Pietsch says, 'Anglerfishes differ radically from all other fishes' (2009, p. 23), and proceeds for several pages to describe 'characters shared by both sexes' (2009, pp. 24–30). However, the properties of males are insufficient to diagnose species. Of anglerfish generally, Pietsch writes, 'With some few exceptions, and despite major efforts, it has not been possible to establish characters that allow intrageneric [species level] identification of males' (2009, p. 24). If we merely caught free swimming male and female specimens of *L. arborifera*, we might never include them in a common category.

As a further blow to the similarity fetishist, the dispositional maneuver does worse here even than it did with ducks. A duck is a separate persistent individual. Even though I despaired of answering the question, it at least seemed *prima facie* coherent to ask what color her head would have been if she had been male. The male anglerfish, which has become a parasitic set of testicles, is not so obviously distinct. To describe the host female as 'a kind of self-fertilizing hermaphrodite' (Pietsch and Kenaley 2007) is already to treat them together as one organism. So it is less clear what object is supposed to have the unrealized disposition *bioluminescent-if-female*.

If we focus attention on the underlying causal process, however, matters are less worrisome. There are females – which is to say that the cluster of properties appears repeatedly in separate individuals – because of ongoing sexual interactions between females and males.

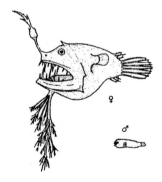

Figure 6.2 Female and male anglerfish of the species *Linophryne arborifera*. (©1982 Tony Ayling, used under the Creative Commons Attribution ShareAlike 1.0 license.)

One might take this just to show that *L. arborifera*, although it is an HPC, is not a natural kind. This would be right if natural kinds were simply those categories which strongly support inductive inference. Indeed, many philosophers take natural kinds to be those that appear in laws of nature or at least strong, law-like regularities (Chapter 1 § B.4). However, such conceptions of natural kinds strain under the weight of similarity fetishism. As we have seen, Boyd thinks that an HPC is defined both by the causal mechanism and by a singular cluster of properties. This leads him to attribute uniform conditional properties to all members of a species, so as to make them all similar. Yet he does not think that taxonomy is simply a matter of finding kinds that will support induction. He writes, 'It is a truism that the philosophical theory of natural kinds is about how classificatory schemes come to contribute to the epistemic reliability of inductive *and explanatory* practices' (1999a, p. 146, my emphasis). Systems of classification are implicated in both induction and explanation, and there is no reason to think that induction is always doing the heavy lifting.

Explanatory considerations identify *L. arborifera* as a legitimate taxon, even if it is not an *inductively* robust category. The fact that populations of *L. arborifera* continue from generation to generation relies on the presence of both males and females. It is not merely that each female right now has the same *type* of causal process maintaining her characteristic features (the physiology characteristic of the species) or even that each female was the result of the same *type* of casual reproductive process (parasitism and so on), although both are true. Rather, each and every member of the species is part of the same *token* causal process – namely, the evolutionary history of the species. This will be true for any lineage, because the members participate in a common causal history over evolutionary time.

One might object to these explanatory concerns in this way: Considering *L. arborifera* and its reproductive parasitism, the females are the autonomous organisms. The explanatory framework most appropriate for thinking about them would treat males as a mere intermediate step in the reproductive process rather than as distinct individuals, analogous to the way we tend to think about sperm in a less dimorphic species. So the females genuinely constitute the important kind.

To press the objection, consider also recently discovered Osedax bone worms. Females root to the carcass of a marine vertebrate, such as a whale, and digest the bone. Males, which are often just one hundredth the size of the female, accumulate in the female's tube. Males do not share observable characteristics with females, for unaided observation at least, because only the females are large enough to see with the naked eye. Surely, the objection has it, Osedax females are the natural kind – they share many properties with one another but few with Osedax males. (This genus of worm was only discovered in 2002, but has already been the subject of much attention. See Rouse et al. 2004, 2008, 2009; Worsaae and Rouse 2010.)

Admittedly, if all or most species were like anglerfish and Osedax, then we might well think of males as a chain in the reproductive process rather than as individuals on par with females. However, biologists recognize these species as oddities. Pietsch enthusiastically portrays anglerfish as 'among the most intriguing of all animals, possessing a host of spectacular morphological, behavioral, and physiological innovations found nowhere else' (2009, p. 8). Worsaae and Rouse describe Osedax as *'strange, bone-eating marine annelids* with "roots" that devour the sunken bones of whale and other vertebrates' (2010, p. 127, my emphasis). So the biology that would be appropriate for a world full of anglerfish and Osedax cannot be a guide to biology for our world. Just as hard cases make bad law, exotica make bad taxonomy.

The crucial point is that explanatory categories are not one-offs. Particular explanations figure in a discipline's overall explanatory strategy. Our attention is drawn to the interaction of *L. arborifera* females and males because of our prior encounters with less wildly polymorphic species. Species taxa are explanatorily important in biology as practiced, and biologists in the business of identifying species identify the HPC.

The approach I have outlined here requires some minimal conditions for how species are identified: For example, a taxon of sexually reproducing organisms must include both the males and the females. Another constraint, which I have simply assumed so far, is that all the members of the species must be organisms. Other products of the species' causal history (such as excreta) will not count as part of the species. These are not arbitrary stipulations, but reflect the explanatory and empirical demands of actual biology.

The requirement that species have a single causal history is not toothless. Carl Craver (2009) argues that the HPC approach is insufficient to pick out determinate kinds. (Craver's notion of a *mechanism* is, I think, more local than something like the natural history of a species. Regardless, his worries about mechanisms can motivate parallel worries about causal histories.) However, many of his arguments are directed at HPCs that are sustained by distinct but relevantly similar causal processes. If we were to think of the kind **economic recession** as an HPC, for example, then we would have to say what makes the causes of different recessions the same kind of cause. No such problem arises for a species, because all its members are the result of a single historical process. Craver also worries that explanatory considerations will be inadequate to distinguish individual mechanisms (2009, § 7). As I argued in Chapter 3 § B.1, scientific considerations are sufficient for at least some specific species taxa to count as natural kinds. All the species I have discussed reproduce sexually, so interbreeding and reproductive isolation criteria can serve to distinguish boundaries. The usual biological considerations are in play. Since this is not a matter to be resolved by *a priori* philosophy, my conclusions are contingent and not final. If the explanatory considerations of biology are insufficient to chart the boundaries of species, then metaphysics is not up to the task either.

The considerations on which I have relied do not require any specific account of *explanation*. We need not think of explanatory power as free floating score that accrues to a theory or system. As Heather Douglas (2009) argues, good explanations should yield predictions. The upshot of the argument I have given is that systematic explanatory prediction, unlike straight induction, requires more than just disparate kinds held together by similarity.

If we see an HPC as an explanatory instrument, rather than as a narrowly inductive one, we shed the vestiges of similarity fetishism. Since considerations of induction form only one aspect of my account of natural kinds, there is no problem with identifying these superficially heterogeneous groups as natural kinds. Identifying them is required for attaining explanatory and predictive success in the domain, and that is just what I take to be the hallmark of natural kinds.

One might worry that this explodes the HPC approach and that these are no longer properly called 'HPCs'. I am not an essentialist about philosophical doctrines, so I would not struggle too much over labels. Nevertheless, I see the view I have defended as an elaboration of Boyd's original proposal. At the risk of being pedantic, let's look at each term in the 'HPC' moniker and ask if it applies:

Are the kinds in my view *homeostatic*? One might worry that homeostasis is now a misleading metaphor, because it is not just a force to produce the *same state*. Yet homeostasis has often been used metaphorically in this broader way. Recall the passage from Eldredge and Gould (§ A.2 above). So I answer yes.

Are the kinds in my view characterized by *properties*? Yes. The approach begins with properties of kind members and charts how these properties are created, sustained, and connected. Note that the properties are not something that could simply be set aside once we had identified the relevant causal history. After all, the history produces many disparate products: *L. arborifera* produces photons which diffuse into shadowy depths, excrement that sinks to the bottom, and so on. These are not themselves members of the species because they have not got the right properties.

Are the properties arranged in *clusters*? On my view the 'property cluster' of an HPC need not be a single list of characteristics which all members of the kind are even likely to possess. It will be whatever complex array of property regularities are sustained by the underlying causal process. These might be a single cluster (for an asexual species), distinct but overlapping clusters (as with ducks and drakes), or almost entirely distinct clusters (as with anglerfish and Osedax). What matters is the complex of related property clusters maintained by an underlying causal process. I see this as a natural development of the original cluster idea: The kind is characterized by patterns of properties among its members, but these do not form a single list of properties which are necessary and sufficient for membership. Indeed, there might be no individual who fully exemplifies all the properties which are maintained by the underlying process.

B. Species and token histories

I have argued so far that species can usefully be thought of as HPCs, but with two corrections to the way that HPCs are usually understood. First, the properties do not appear on a single unranked list. The properties may have more structure than that, and in polymorphic species will form a complex of related clusters. Second, the local causal mechanisms that maintain the species – the physiology, development, and reproduction – are recurring parts of a single overarching causal history. The so-called homeostatic process which maintains a species is unified by being, at the largest scale, a single token history rather than a type of causal process that can occur independently at separate times and in separate places.

In the next several sections, I develop the latter point in greater detail. In § B.1, I consider philosophers' thought experiments which are meant to show that species need not be united in this way. In § B.2, I consider more biologically plausible cases of a species arising more than once.

B.1 The tigers of Mars

When astronomers wonder whether there is water under the surface of Mars, there is no problem thinking that the stuff there might count as water. The kind **water** is not an HPC, at least not in the sense that it is all the result of some shared causal history. Water which formed from hydrogen and oxygen on a distant planet is just as much water as H_2O here on Earth. (For more about water, see § E below.)

Essentialists about natural kinds presume that members of any kind are similar, united merely by shared intrinsic features. (Some of the discussion here echoes my reasons for rejecting the intrinsicness assumption, Chapter 1 § B.8.) So essentialists often suppose that biological categories would behave on other planets just as chemical categories (like **water**) do. For example, Hilary Putnam asks us to imagine a case in which 'we discover "tigers" on Mars. That is, they look just like tigers, but they have a silicon-based chemistry instead of a carbon-based chemistry.' He asks, 'Are Martian "tigers" tigers?' and answers, 'It depends on context' (1975*b*, pp. 157–8). If context directs us to observable features, then the silicon doppelgangers count as tigers on account of their stripes and fearful symmetry. If context directs us to genetic and chemical features, then they do not count as tigers on account of their different composition.

We might imagine a slightly different scenario, in which either context would lead to the same judgment. Imagine tiger-like quadrupeds on Mars that do not merely *look* like tigers but which are also molecule-for-molecule duplicates of tigers. Philosophical scenarios of this kind are often posed in terms of *Swampman*, a random ensemble of matter risen from the swamp which is identical in chemical composition to Donald Davidson (Davidson 1987). Imagine the

Marstiger, in a similar way, to be an ensemble of matter risen from behind a Martian rock which is identical in chemical composition to some specific tiger. Even if we would call Marstiger a "tiger", there are legitimate reasons not to count it as a member of the species *Panthera tigris*. The historical and causal process which explains why members of the species are similarly tiger-like does *not* explain why Marstiger is tiger-like. Since Marstiger is genuinely a random product, like Swampman, then there is no explanation – its tiger-likeness is just a curious but inexplicable outcome. If instead Marstiger were the result of some complicated evolutionary history on Mars, then it would be that history which explained its structure. The parallels between its Martian species and *Panthera tigris* would be merely coincidence, the result of chance similarities between their evolutionary environments.

T. E. Wilkerson offers a slightly less abstruse thought experiment. Imagine, he suggests, 'after a series of mutations, gorillas and humans might evolve into animals that are very similar – as similar as, say modern Europeans and modern Chinese. Some of the animals would be descended from gorillas, some from humans, yet it is far from obvious that we should regard them as belonging to different species' (1995, pp. 112–13). This scenario is less mind-bending, perhaps, because it begins with two similar kinds of creatures (gorillas and humans, rather than tigers and Martian fauna) and it tells a story about them becoming similar (rather than identical). Nevertheless, Wilkerson acknowledges, 'Those who prefer philosophical fables about Twin Earth can construct a similar story' (1995, p. 113). What should we say about Wilkerson's scenario? If we imagine the scenario continuing, so that the descendants of humans and descendants of gorillas intermingle and form mixed populations, then they would form an HPC. The persistence and reproduction of these future humanoids – or, as we might just as well say, future gorilloids – would result from the causal cohesiveness of the combined population. The group would not be monophyletic, because it would have two distinct ancestor species, but it would count as a single species by common biological criteria. The organisms would be interbreeding, and natural selection could operate on them as an evolutionary unit.

Since Wilkerson is trying to construct an example in which an intuitive criterion of similarity disagrees with those biological criteria, we should imagine instead that the two populations are separate and do not interact. We should imagine also that the resemblance of the future humanoids (the descendants of humans) and gorilloids (the descendants of gorillas) is not the result of some common influence in their evolutionary development. They are similar, but just as a matter of coincidence. Consider one feature they have in common – imagine, say, that both evolved transparent brain cases. The explanation for why future humanoids have transparent brain cases will rely on the evolutionary path that led present humans to have distant descendants with transparent brain cases, and the prediction that offspring of future humanoids will have transparent brain cases will depend on reproductive and environmental facts about them. Contrariwise, the fact that future gorilloids have transparent brain cases would not be explained by the history leading forward from humans but instead by the evolutionary history starting with gorillas. What matters is the causal history. Since there are separate histories in the scenario, there are separate species. This amounts to thinking of each species as an HPC.

In the case Wilkerson imagines, there could perhaps be a *further* natural kind that includes both the humanoids and the gorilloids, since they are stipulated to be very similar. The point is just that it would not be an HPC. The thought experiments about Marstiger or future gorilloids do not undercut the systematic idea that some important natural kinds are HPCs, and that understanding the natural kind structure in a domain requires recognizing HPCs where they occur.

B.2 Hybrids and separate origin

Ruth Garrett Millikan (1999) defends a version of the HPC view which, like mine, requires historical and causal connectedness for species membership. She has similar things to say about the Swampman thought experiment (Millikan 1996). My differences with Millikan on this are at the level of nuance. She writes, 'Biological kinds are defined by reference to historical relations among the members, not, in the first instance, by reference to properties' (1999, p. 54). I worry about what *the first instance* is meant to be. As I argued above, discovering an HPC may start with recognizing a property cluster. Also, the structure of properties is an important part of the HPC.

Boyd, in a commentary on Millikan's article, resists more strongly. He flatly denies 'that HPC kinds are defined by reference to historical relations among the members, rather than by reference to their shared properties' (cited above). Nevertheless, he still insists that 'biological species are necessarily limited to a particular historical situation' (Boyd 1999c, p. 80). These positions are *prima facie* inconsistent – either species are historical entities or they are not!

The inconsistency is resolved by considering the kind of case Boyd has in mind. He does not mean to open the door to philosophers' examples like Marstiger. Rather, he thinks that species can arise by hybridization and that crosses may occur separately. He writes, 'there are actual cases of recognized and evolutionarily stable plant species which arose from hybridization ... [T]he stability of such a hybrid species over time will typically (perhaps always) be partly explained by the sorts of historical relations Millikan has in mind *among members of the "parent" species*. But all this is compatible with the independent evolution of two or more lineages within the hybrid species' (1999c, pp. 80–1, emphasis original).

Although Boyd considers the case of plant species, a similar point can be made with animal hybrids. Take the example of the Mariana Mallard, *Anas oustaleti*. It was once common in the Mariana Islands but is now extinct. It was a stabilized hybrid of the mallard (*A. platyrhynchos*) and the Pacific Black Duck (*A. superciliosa*). Imagine we were to discover a population of ducks outside the Mariana Islands which had originated at about the same time as the Mariana Mallard from a separate hybridization, in a similar environment, between a mallard and a Pacific Black Duck. It might be empirically impossible to distinguish such a population from a wayward, surviving group of Mariana Mallards. That is just an epistemic point, however. Supposing that a species must be a monophyletic group (in the strict sense that the members of a species must be all and only the descendants of some determinate ancestors) then the other hybrid ducks would not be members of *A. oustaleti*, even if there was no way to tell the difference.

Philip Kitcher (1984) uses examples of hybridization, like this one, to argue that species should be distinguished by *structural similarity* rather than by *phylogenetic relationships*. Michael Devitt (2008, pp. 346–7, fn. 6) makes a similar move *en passant*. Given this dilemma, mere similarity should win in some cases. For example, consider an alternative scenario in which mallards and Pacific Black Ducks are regularly hybridized for commercial purposes. The resulting hybrid ducks would, when allowed to interbreed, form small monophyletic groups – but the whole commercial population, taken together, would be a polyphyletic group. Yet the commercial population would exhibit a stable cluster of properties, it would support induction and explanation, and so on. It would thus be a natural kind. (Recall that being a *natural kind* in this sense is compatible with being artificial; see Chapter 1 § B.5.)

Nevertheless, Kitcher poses a false dilemma. Accepting species concept pluralism, we can insist that the phylogenetic species concept and the requirement of monophyly are not always appropriate (see Chapter 3 § B), without insisting that mere structural similarity is ever enough. Mere similarity would allow Marstiger to be a conspecific of tigers, even though there is no stable basis for induction or ground for explanation. For the commercial ducks, there would be more than *mere* similarity. There would be a complex causal process involving the distinct parent species with their complex evolutionary histories plus an agricultural practice repeatedly hybridizing them. So the commercial ducks would form an HPC kind, with a unified undergirding causal story. This is not an especially exotic thought experiment. Parallel practices are common with plants.

Now return to the example of hybridization in the wild. Boyd insists that separate hybrid lineages might comprise a common HPC, even though the lineages do not have a common origin. One way of understanding this is to say that – although the two do not share the same *token* causal origin – they originated in the same *type* of way. To illustrate the difference, consider an example suggested to me by Richard Samuels: the phenomenon of sunburn. If several people get sunburn at the beach, their symptoms are caused by the very same causal agent; namely, the sun. If one person gets sunburn at the beach and another gets sunburn from lying under a tanning lamp, their symptoms are caused by different things; one by the sun, the other by the heat lamp. In all the cases, sunburn is caused by ultraviolet radiation. In the former cases, the cause that produces the UV is the very same token cause (the sun). In the latter cases, there are two different token causes of the same type (**UV source**). As I developed the HPC approach, above, members of a species had to result from the same token causal process (the evolutionary history of the species). The commercial duck hybrids similarly would be the result of one token process (the biological–agricultural complex). In order to allow distinct hybrid lineages in the wild to be one species, we might relax the requirement so as to only require the same type of cause.

Let's call an HPC unified by a single token causal history a *token-HPC* and call an HPC unified only by the same type of causal histories a *type-HPC*. The suggestion is that (some) species are type-HPC kinds. If similar type and token identical processes were the only option, then this might be the best way to understand hybrids in the wild. However, like Kitcher's opposition between mere similarity and monophyly, it is a false dilemma. Even if we accept separate hybrid lineages to be members of the same species, structurally similar causal processes on a planet in the vicinity of Alpha Centauri could not – even if similar down to the molecular level – produce a Mariana Mallard. The hybrid species result from a cross between members of the very same parent species, rather than a cross between merely similar organisms. For the case of *A. oustaleti*, the two parent species are token-HPC kinds. So the type of process which could produce a member of the species requires historical connection by way of the parent species.

We can imagine more exotic cases in which a hybrid species results from separate hybridization events between parent species which are themselves hybrid species, and so push the historical connection even further back. In such a case of long-separated common origin, the resulting hybrids might be so diverse as to not form a natural kind. If they do form a unified kind, resulting from disparate causal processes, the natural history must involve token-HPC species somewhere in the past. Even if we can imagine a planet with two parallel but never crossing streams of life, Earth is not such a planet. So hybrids will never be *mere* type-HPCs, connected only by the *similarity* rather than the identity of the causes that produced them. A *hybrid-HPC*, unlike a token-HPC, results from separate causal histories; yet, unlike a type-HPC, those histories are necessarily limited by historical connection to some token-HPCs.

As an aside, I do not have a strong opinion on whether or not separate hybrid lineages could count as members of the same species. My point is, rather, that the possibility has tempted philosophers like Boyd and Kitcher to deny that members of a species must share a common causal history. If one feels a similar temptation to deny that species must be token-HPCs, one ought not to relax the requirement so much that species might be mere type-HPCs; we should require that they be at least hybrid-HPCs.

Once we have marked these distinctions, a crucial question for the HPC approach is whether any natural kinds will turn out to be type-HPCs. I will return to this in § E. Between here and there, I want to deal with the relationship between the HPC approach and the species problem.

C. RE: ducks, the species problem redux

As we saw in Chapter 3 § B, the 'species problem' is a label for three distinct issues. First, there is the question of whether species taxa like **mallard** are real features of the world. (I argued that at least some species taxa are natural kinds.) Second, there is the issue of whether the category **species** is a feature of the world. (I argued that the category is not a natural kind.) Third, there is the question of whether particular species are individuals or classes. (I did not address this issue in Chapter 3, but I will turn to it in the next section.)

For specific taxa, we saw the examples of the species complex *Anopheles gambiae* sensu lato, the species *A. gambiae* sensu stricto, and the incipient species *A. gambiae* M form. I argued that each is a natural kind. Given the historical and reproductive unity of these groups, we can now add that each is a token-HPC. Cases from this chapter, like *Anas platyrhynchos* and *Linophryne arborifera*, also illustrate specific species taxa which are natural kinds and token-HPCs. Indeed, the strategy for charting HPCs (§ A.2) gives us a playbook for resolving the first aspect of the species problem in particular cases. It is possible to show that taxa are natural kinds without describing them as HPCs, as I did in Chapter 3, but the account of HPCs fortifies and illuminates the general account of natural kinds.

For the **species** category, we saw that it was undercut by two kinds of pluralism: category pluralism (the fact that species might be distinguished on the basis of different reproductive, ecological, or evolutionary criteria) and rank pluralism (the fact that species might be charted at a larger or smaller scale). Robert Wilson suggests that we can accommodate both forms of pluralism by treating the category **species** itself as an HPC. That is, he takes the playbook for species taxa and tries to run plays for the category. He sees this move as motivated by the same sentiment as the HPC approach itself. The approach, he maintains, is 'integrationist or unificationist about natural kinds' (Wilson 2005, p. 116). This formulation of the sentiment is correct in one sense, namely that each *particular* HPC is unified and integrated. Wilson requires the further sense that HPC kinds in general are unified and integrated with one another, but that simply is not so. Separate HPCs might be maintained by diverse and disjoint causal processes. Whether the **species** category can viably be seen as an HPC depends on the details. Wilson argues the details thus:

> The HPC view can ... be applied to the species category, allowing a definition of what sorts of thing a species is that marks it off from other biological categories. First, the general nature of the cluster of properties – morphology, genetics, genealogy, and so on – will distinguish species from nonevolutionary natural kinds such as cells (in physiology), predators (in ecology), and diseases (in epidemiology). Second, species will be distinguished from other evolutionary ranks ... by the particular specifications of the general cluster of properties. (2005, p. 111)

There are three problems with this approach which together are enough to put the kibosh on it.

First, the structure of the kind **species** is very different than the structure of particular biological taxa. The members of the species *A. gambiae* are individual organisms, and so are the members of the incipient species *A. gambiae* M form. The property structure of the wider kind will be different than the property structure of the narrower kind, but the same sort of properties will appear in each: e.g. morphological, behavioral, reproductive properties which tend to be possessed by individual members of the kind. Yet the members of the kind **species** are taxa. The structure which characterizes it will not be a structure of the properties of organisms, but of structures characteristic of taxa. Treating taxa as first-order kinds, **species** is a second-order kind. With individual taxa, it is easy to see how causal factors in organisms might underwrite the tendencies that sustain the structure of properties. With the **species** category, it is less clear how a homeostatic mechanism could maintain the requisite *structure of structures*.

Second, it is unclear how construing the category **species** as an HPC helps with any of the problems raised by pluralism. Regarding category pluralism, take a case where reproductive and ecological criteria yield different answers as to which taxa are species. Both reproductive and ecological features are part of the cluster taken to define **species**, so we have a boundary case – but that is just to say that we have no more answer after than we did before construing **species** as an HPC. Regarding rank pluralism, consider again the example of *A. gambiae*. There was no strong reason to say that one of the species complex, the species so-called, or the incipient species is genuinely the *species*. The HPC definition gives us no further guidance. Wilson admits as much, but insists that it is a merit of his approach: 'In some cases, the distinction of species as a particular rank in the biological hierarchy will be difficult to draw, but I suggest that this is a virtue ... of the HPC view, since varieties sometimes are very like species (for example, in the case of ... incipient species), and species are sometimes very like genera' (2005, p. 111). If **species** were a natural kind, then of course we would not want our categories to be any sharper than the kind in the world. (I argued for this in rejecting the sharpness assumption; Chapter 1 § B.6.) Yet the problem of rank pluralism is rather general. Although it may only be salient 'in some cases', that is just because context and our interests lead us to specify a species taxon in other cases. This comports with saying that, although species taxa are natural kinds, the category is not.

In Wilson's defense, one might reiterate what he offers as the first important function of the **species** category: It distinguishes species taxa from kinds like **cell** and **predator**. However, Wilson does not distinguish token-HPCs from type-HPCs. Species taxa, as we have seen, are token- or hybrid-HPCs. In contrast, **cell** and **predator** – if they are HPCs at all – are type-HPCs. There need be no historical connection between two predators. Marstiger might be a predator, if it lives in an ecosystem and lives by hunting, although its causal isolation from Earth means that it cannot possibly be a tiger. So the categories **token-HPC** and **type-HPC** suffice to do the work of distinguishing biological taxa from other, non-historical kinds. Note also that Wilson acknowledges the enquiry relativity of natural kinds in passing, by noting that **cell** is a natural kind *in physiology* and so on. The difference between species taxa and **cell** might, in part, result from the different but overlapping domains of objects and phenomena that the kinds are meant to chart.

Third, if the **species** category is an HPC, it can only be a type-HPC. As such, even if we could make sense of what the homeostatic process was supposed to be, we would have to further specify how to identify causal processes of that type. Two separate species might evolve under separate conditions. All life on Earth shares some causal origin in the distant past, but that minimal common descent does not do much to distinguish species from other taxa and from arbitrary sets of organisms. Moreover, it seems reasonable to think that extraterrestrial organisms might be organized into species taxa. Settling that depends in part on how we understand **life**, an issue which probably cannot be resolved in an enquiry-neutral way; see Chapter 2 § B.2. Although I have argued that species taxa as token- or hybrid-HPCs are natural kinds, we should still worry about the *bona fides* of type-HPCs. I return to that general problem in § E.

To sum up, Wilson's stratagem of construing the **species** category as an HPC faces serious difficulties. So I suggest that understanding species as HPCs underscores, rather than undercuts, the conclusions about the first two aspects of the species problem which I reached in Chapter 3. This leaves the third aspect of the species problem, *individualism*. I finally turn to it in the next section.

D. Historical individuals

A widely accepted view about species – both among biologists and philosophers – is that species are individuals rather than kinds. This is called the *individuality thesis* or *individualism* about species. The view was originally proposed by Michael Ghiselin (1966, 1974) and notably defended by David Hull (1976, 1978). Ghiselin has, with some justification, taken to calling it the philosophical consensus (2002, p. 153, 2007, p. 283, 2009, p. 253). Even philosophers opposed to individualism acknowledge it as the default position; e.g. Keller, Boyd, and Wheeler (2003).

Ghiselin takes individualism to have far-reaching consequences. 'It provides,' he writes, 'the inspiration for a new ontology with profound implications for knowledge in general' (2009, p. 253). It raises two concerns about my project. First, individualism is sometimes taken to be a reason to reject the idea, which I have defended, that a species is an HPC; HPC kinds are taken to be sets rather than individuals. Second, individualism threatens the whole project of thinking about science as charting natural kinds; the business of many sciences, if Ghiselin is correct, is to chart something *else*.

There are three general kinds of arguments given for individualism. First, it is taken to better reflect the language used to describe species; species are treated grammatically as individuals rather than as predicates. Second, individualism is taken to better capture the structure of species; sets and classes are argued not to have the right kind of structure. Third, it is argued that individuals (but not sets) are capable of change. I will not provide much attention to the first kind of argument, which turns on the grammar of species talk. A metaphysical point which turns on the use of prepositions is too abstruse to be worth arguing about. Moreover, the approach to natural kinds that I have argued for takes them to be primarily a matter of taxonomy rather than nomenclature – structure in the world rather than the usage of words (see Chapter 1). I will deal instead with the latter two arguments, starting with the former.

D.1 Sets and sums, a metaphysical non sequitur

There are two formal ways of thinking about collections or aggregates, things which are put together out of other things. One approach is *set theory*, which treats a collection as a class of objects such that the things that go into it are *members* of the class. The other approach is *mereology*, which treats a collection as a composite ensemble such that the things that go into it are *parts* of the ensemble. It is tempting to treat the debate about individualism as a debate about which of these (set theory or mereology) is appropriate for thinking about species. I will ultimately argue that this is a mistake, but it will help to be clearer about the difference between the two.

Set theory begins with some initial collection of individual objects. A *set* is a collection of objects without regard to their order. A set can itself be treated as an object, so a set can be a member of other sets. For example, the set containing all dogs and cats contains a large number of animals; the set containing the set of all dogs and the set of all cats contains just two things, both of which are sets rather than animals. Because order does not matter, you cannot make a different set by changing the order of objects in it. To give you an image: You can think of a set as being a metaphysical wrapper that extends around all of its members. That package can be put in another package to make a different set. Of course, set membership is not a physical wrapper. Although there are a computer and a coffee cup on my desk, there is not *a set containing the computer and the coffee cup* on my desk. The set is an abstract object, in the sense that it does not have a physical location.

Whereas the basic relation in set theory is membership, the basic relation in mereology is the relation between part of something and the whole thing. Mereology is concerned only with individuals, although the individuals may be composite. For example, the front half of the family dog is part of the dog. The dog is composed of its front and back halves; in the language of mereology, the whole dog is the *sum* or *fusion* of its two halves. Just as set membership is not a physical wrapper, mereological fusion is not a physical operation. Saying that the dog is the fusion of its front and back halves is just to say that it is a whole composed of those parts; it is just as true to say that it is the fusion of its left and right halves.

The debate about species individualism, understood as a metaphysical debate about whether organisms are *members* or *parts* of a species, turns on whether species are sets or fusions – whether they are the proper objects of set theory or mereology. Philip Kitcher argues, against individualism, that any constraint of historical continuity which can be satisfied by parts of an individual might similarly be satisfied by members of a set. One need only insist that 'species are special kinds of sets (namely historically connected sets)' (Kitcher 1984, p. 314).

Berit Brogaard argues, contra Kitcher, that sets are unfit to be species. Considering a set that contains the organisms that are part of a species, she writes, the set '*represents* in certain ways the concrete aggregate of organisms under consideration but it is not identical to it, and does not coincide with it' and 'the members of sets exist within the set without order or location' (2004, pp. 229, 230). The set strips members of their history, she insists, so putting the organisms in a set severs the requisite historical connection. Similar rhetoric is offered by Michael Ghiselin, who insists: 'An individual is a single thing, definitely located in space and time. It is integrated in one way or another – joined as by physical or social forces or common descent, whereas the members of a class need share only traits' (1981, p. 271).

Their argument can be illustrated with the simple case of a four-piece jigsaw puzzle. Suppose we put the pieces together so as to complete the puzzle, as in Figure 6.3a. We can think of the assembled puzzle as a composite individual, as in Figure 6.3b. This composite is the fusion of the four puzzle pieces. We can also think of the puzzle as a set, the members of which are the four pieces. If we write down the set in the usual way, we list the set members with commas between them. This yields Figure 6.3c. Looking at the set written down in this way, it is tempting to say that the puzzle has literally been taken apart. The fusion, by leaving the pieces where they are, preserves the fact that the puzzle is solved.

The argument, so construed, relies on some representational legerdemain. The set recorded in Figure 6.3c separates the members of the set only because that is the conventional way to write down a set. The set is an abstract object, and mere set membership does not involve spatially arranging or causally manipulating the pieces. Yet the set members, the pieces themselves, are concrete objects. They do have determinate locations and relations to one another – exactly the same relations they have if we consider them as individuals. Brogaard is right that set members do not have order or location *merely in virtue of being members of some set*, but the set does not somehow eradicate the order and location they do have. We can consider sets or classes of things which are spatiotemporally unrestricted, but that does not mean that all sets are spatiotemporally unrestricted. The set of puzzle pieces is a set of things in determinate places. So Kitcher is right that the metaphysics of species as individuals might be reconstrued in terms of suitably specified sets.

If one were sympathetic to individualism, one might still insist that it makes the causal and historical cohesion of species more salient than thinking in terms of sets does. For example, Richard Richards argues that individualism can 'give all sorts of guidance in thinking about species', that it 'promises to be a fertile way to think about species within the evolutionary context' (2010, pp. 175–6). I agree that species are best understood by paying attention to the cohesive forces which unify them over time. Yet there are at least three reasons why this fails to establish individualism.

First, as a formal matter, the parts of a mereological fusion are not restricted in time or space. Kitcher offers the example of 'a single organism whose life consists of the last decade of Bertrand Russell followed by the first decade of the family dog.' He is tempted to say that such a gerrymandered thing would not be 'a natural object', but I think that much of this temptation arises from the fact that the fusion of late Bertrand Russell and early Bertie the dog does not constitute an *organism*. It lacks the right kind of causally connected history. As a matter of metaphysics, however, it is less obvious that the fusion cannot be treated as a thing. The general logic of

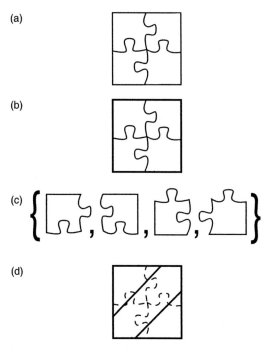

Figure 6.3 (a) Take, as an example, a simple four-piece jigsaw puzzle. (b) The mereological sum or fusion of the four pieces is a square, and each piece is a part of it. (c) The set containing the four pieces is an abstract object. (d) The fusion also has parts that cut across the puzzle pieces. For example, it has three parts that are the diagonal strips as shown. These three diagonal parts constitute the puzzle.

mereology allows us to do so. To take less abstruse examples, we do in fact acknowledge some individuals which have spatially separated parts. Kitcher gives the example of a broken chair (which can be one object even though its parts are on opposite edges of the work bench) and a nation (which can have spatially separated territories) (2001, pp. 46–7). So species being individuals does not guarantee that their parts will be spatiotemporally connected. Rather, the account must be that species are *particular kinds* of individuals – just as species being classes requires that they be *particular kinds* of sets.

Second, a mereological fusion does not preserve a record of the parts which were fused to compose it. This, in fact, is a primary difference between mereological thinking about individuals and set theoretic thinking about classes. As Henry Leonard and Nelson Goodman explain, although a fusion is formally specified as a collection of parts, 'to conceive a segment as a whole or individual offers no suggestion as to what these subdivisions, if any, must be, whereas to conceive a segment as a class imposes a definite scheme of subdivision – into subclasses and members' (1940, p. 45). So the jigsaw puzzle is the fusion of the four puzzle pieces, but it is also the fusion of three diagonal parts that crosscut the puzzle pieces (see Figure 6.3d). Of course, this is true even without there being physical cuts across the puzzle. Just as treating the pieces as members of a set was merely a formal point about the unchanged puzzle pieces, noting that the fusion of the four puzzle pieces is the same as the fusion of the three diagonal pieces is merely a formal point. Nevertheless, it highlights a heuristic *weakness* of individualism. Conceiving of a species as a suitably specified set of organisms gives us ready access to information about the number of organisms that are members of the species; it is the cardinality of the set. Conceiving of a species as an individual makes numerical information less salient.

Third, the *part of* relation is transitive. To take a mundane example: A particular fuse is part of my car's electrical system, and the electrical system is part of the car. It follows that the fuse is part of the car. Similar logic does not hold for species. The family dog's left leg is part of the dog. If the dog is part of the species, then it follows that the dog's leg is also part of the species. This is an odd thing to say. Resisting it requires saying either that a dog is not part of a species in the same sense that the leg is part of the dog or saying that mereology, the logic of parts and wholes, breaks down in this case. Of course, the point can be finessed by distinguishing organisms as privileged or important parts of species. No such gymnastics are required if species are sets, because set membership is not transitive.

The first objection suggests that species-as-sets and individualism are on equal footing, while the latter two objections reveal advantages of species-as-sets. I do not take any of these reasons to be decisive. The point becomes one of deeper metaphysics.

One might consider sets to be somewhat mysterious, because they do not exist in space and are formal rather than causal entities. One might prefer individualism over a set-theoretic alternative for this reason. It would cease to be a point about species, however, and become simply one consequence of metaphysical nominalism. Indeed, Goodman and others developed mereology in their attempt to banish abstract entities entirely. I do not have strong opinions about such a crusade. It is concerned with different questions than the ones I am addressing here. By arguing for the existence of some natural kinds, I deny the kind of nominalism which claims that natural kinds are no different than any arbitrary collection of things. I have no beef with nominalism which can recognize the difference between natural kinds and other kinds but which tries to do so without abstracta.

D.2   Metaphysical puzzles about change

There is a separate line of argument for individualism, motivated by the fact that species can change. Given evolution, any view of species must view them as historical entities. Sets, however, are abstract objects and so cannot change.

Here again, I think the debate wades too deep into metaphysics. Considered with sufficient scrutiny, any account has a hard time explaining change. Individualism is ambiguous as to whether a species should be understood as a momentary individual existing in space or a temporally extended individual stretching across both space and time (cf. Reydon 2008). Call the former approach three-dimensionalist (3-D) and the latter four-dimensionalist (4-D).

For 3-D individualism, *A. platyrhynchos* is the sum of the bodies of the ducks and drakes existing *now*. The species tomorrow will be the sum of the bodies of the ducks and drakes existing *tomorrow*. In terms of mereology, these sums are simply different individuals. So one might object that the present individual will not have changed; instead, it will have been replaced with a different individual. This is the same problem of identity across time that arises in other areas of philosophy. I am not arguing that it is insoluble, just noting that it arises for the 3-D individualist. Change is a metaphysical conundrum if the individual is identified in terms of momentary time slices, because it is not obvious what metaphysical glue holds the momentary individuals together across time.

One way of resolving the conundrum is to treat individuals as 4-D objects. Instead of merely being the sum of *present* ducks and drakes, *A. platyrhynchos* would be the sum of all ducks and drakes that ever were or ever will be. Such a sum is sometimes called a spacetime worm. A common objection to such a view is that spacetime worms (allegedly) do not change. Because it would be the whole history of the species summed up, the 4-D species includes all ducks everywhen as its parts. The parts that are located at Woodlake on December 9, 2011 simply exist there in an unchanging way. Of course, an enthusiast for the 4-D view will insist that this misunderstands change. Change is not about coming into and going out of existence, the 4-D individualist will say, but instead it is simply a matter of having temporal parts with different properties. Changing location, for example, is having a temporal part at Woodlake on December 9, 2011 at 1:30 p.m. and a different temporal part across the street at 1:31 p.m.

This discussion becomes rather abstruse, but note simply that change is a metaphysical hot potato for individualism. Some latitude and elaboration is required to make sense of it. If we allow similar latitude for elaboration in terms of sets, the two ontologies remain equipollent. We might treat the relevant classes *intensionally*, so that they are specified in terms of conditions for membership rather than in terms of specific members (Reydon 2003). Often intensions are modeled as structures of sets or functions from possible circumstances onto sets. Alternately, we might consider the set of all members of the species from across time – a possibility which exactly parallels 4-D individualism (Kitcher 1984, p. 311).

Of course, many people do not find change especially puzzling. Insofar as it is a problem for species as sets, however, it is a general problem about the metaphysics of time. Such general puzzles will not break the tie between the alternative ontologies of species.

D.3 Individualism and HPCs

Ultimately, it is unfortunate that the debate about species individualism has been framed by vocabulary of individuals pitted against sets and classes. Although he uses the terminology of individuals and classes, David Hull is quite clear that the central point is that members of a species must be historically connected and not something about the deep metaphysics of aggregates. He writes:

> By 'individuals' I mean spatiotemporally localized cohesive and continuous entities (historical entities). By 'classes' I intend spatiotemporal unrestricted classes, the sorts of things which can function in traditionally-defined laws of nature. ... The terms used

to mark this distinction are not important, the distinction is. For example, one might distinguish two sorts of sets: those that are defined in terms of a spatiotemporal relation to a spatiotemporally localized focus and those that are not. (1978, p. 336)

His use of contentious vocabulary, stipulated to track a somewhat different distinction, means that others often misread him. This is just a point of rhetoric, however, and not a point of substance. It is the distinction and not the labels that matter.

This important core of individualism is correct, and it is entirely compatible with the view which I have argued for – namely that species are HPCs and natural kinds. As I have argued at greater length in another context, historical individuals may simply be understood as token-HPCs (Magnus 2012). For example, a single individual duck is a stable cluster of duck activity over the course of her life, maintained by processes of development and physiology. All of her *duckiness* is connected as part of her ongoing personal history. As a member of the species, there is little difference between seeing her as part of the aggregate species-individual *A. platyrhynchos* and seeing the cluster of properties that she realizes as an instance of the overall property structure of the species. Either description recognizes the historical and causal unity of the species. As Boyd urges, 'by seeing the similarities between the inductive and explanatory rôles played by reference to natural kinds, on the one hand, and by reference to individuals, on the other, we can see why the distinction between natural kinds and (natural) individuals is ... merely pragmatic.' The distinction is, he suggests, 'almost just one of syntax' (1999a, pp. 163, 164). Boyd is right, the distinction collapses – so long as we are considering only token-HPCs.

E. HPC thinking beyond token causes

As we saw in Chapter 4 § B.2, Anjan Chakravartty defends an ontology of properties clustered together; to use his phrase, properties that are 'sociable'. Although he thinks that HPCs are one kind of property sociability, he argues that they cannot be metaphysically basic. He writes:

The idea of homeostatic clustering ... is certainly attractive, but sociability will not always be analysable in this way. For example,

> it would seem that homeostatic mechanisms are not responsible for the co-instantiation of the mass, charge, and spin of electrons. In the case of many essence kinds, sociability is a brute fact, admitting of no causal decomposition. The presence of homeostatic mechanisms is not the *sine qua non* of kindhood. It is a special case of sociability. (2007, p. 171)

Since I am not trying to defend any particular account of the base-level metaphysics of things, I will not say anything either *pro* or *contra* property sociability as a basic feature. Set that aside. The point, for us, is that Chakravartty is led to posit brute sociability because some regularities simply are not causally maintained patterns of properties in the way that species are. Unlike protons, which are composed of quarks, electrons are believed to be fundamental and indivisible. There is no causal process operating on the universe to make it the case that all the electrons have the same charge, mass, and spin, nor are there processes chugging along inside each electron that is responsible for its properties. Yet this does not preclude **electron** from being a natural kind.

The thing which makes species especially good examples for the HPC approach, I have argued, is that the members of a particular species are all connected to the causal history of the species. As such, although biologists might argue about the limits of a species, they do so only regarding exactly how far the species boundary extends. Anything which is not connected by a train of biological causes to some arbitrary specimen of the species will not count as a species member. Species are what I have called *token-HPCs* or *hybrid-HPCs*; § B.2 above.

Many natural kinds are, if they are HPCs at all, only *type-HPCs*. That is, there is not a single causal nexus supporting the structure of properties. Instead, each instance of the kind has its own separate causal history. A type-HPC might have instances anywhere in the universe, so the task of identifying it and charting its boundaries is more complicated. One must not only determine how far a single instance extends – as one must with any HPC – but also which things in the universe count as instances of the kind.

E.1 The waters of Mars

For example, water formed on Mars probably has no causal connection to water formed on Earth. All that is required is that oxygen and hydrogen atoms meet under the right circumstances. Nevertheless, it has been suggested that water is an HPC kind. Water is picked out by a cluster of macroscopic properties: It is clear, odorless, potable, and so on. These qualities occur regularly together in the same liquid – as Chakravartty would say, the properties of water are sociable. Yet, water is H_2O. The stable underlying chemical structure is responsible for the sociability. Hilary Kornblith argues that the powers of H_2O explain both which properties are in the cluster and which are excluded. He writes,

> [W]hen molecules of hydrogen and oxygen combine to form the stable compound H_2O, the observable properties of being colorless, odorless, tasteless and so on, are an inevitable product of that base. Certain other properties of the compound, however, are not thereby determined. The weight and shape of a sample of H_2O, for example, are not determined by the fact of its chemical composition, nor is its temperature. (1993, p. 37)

Moreover, he insists that 'the chemical bond between hydrogen and oxygen makes H_2O a homeostatic unit' (1993, p. 37). On his view, water has a cluster of properties, and the unity of the cluster is explained by an underlying causal mechanism. In short, water is an HPC.

One might object that the essentialist tradition already identified water as H_2O. Since Putnam, the semi-formal inscription 'water = H_2O' has been the stock example of a metaphysically necessary truth about the essence of something. The HPC approach is typically offered as a rival to essentialism, but here it looks as if Kornblith is just dressing up the essentialist account in HPC clothing and colors.

The objection works against Kornblith, perhaps, but only because Kornblith just discusses water as a stand in for chemical kinds generally. It is just an illustration of a general formula, which he expresses just before the passage cited above: 'When certain unobservable properties reside in a homeostatic relationship a collection of observable properties inevitably flow from that unobservable base' (1993, p. 37). Water, used as a passing example of a chemical kind in this way, is portrayed as a static aggregate of H_2O molecules. Just as there are a few dozen gherkins in a large jar of pickles, one thinks of a few dozen moles of H_2O in a liter bottle of water.

Importantly, the actual microstructure of water is more complicated. Robin Hendry explains that 'macroscopic bodies of water are complex and dynamic congeries of different molecular species, in which there is a constant dissociation of individual molecules, reassociation of ions, and formation, growth, and dissociation of oligomers. Being H_2O, if understood as a molecular condition, cannot capture the molecular complexity of water' (2006, p. 870). For one thing, two H_2O molecules may dissociate into an H_3O^+ ion and an OH^- ion. Predictable quantities of these ions will be present in any sample of liquid water, with quantities varying as a function of temperature and pressure. If only H_2O molecules counted as instances of the kind **water** – and thus if H_3O^+ and OH^- ions did not – then dissociation would be a matter of breaking down into impurities. As a consequence, pure water would become impure as a matter of chemical inevitability. It makes the most sense to say instead that the ions are part of the water, and so something besides H_2O can be part of a mass of pure water. For another thing, H_2O molecules form larger structures by clustering and hydrogen bonding. These larger-scale structures are responsible for water's anomalous features. So the water in the bottle is both less and more than just H_2O. (Martin Chaplin (2010) provides an excellent guide to the chemistry of water.)

Details like these lead Paul Needham to deny that the kind **water** is distinguished by microscopic features at all. Rather, he suggests, 'macroscopic criteria determine sameness of substance kind, whose variable microstructure is then made the subject of scientific investigation' (2000, p. 20). There is a sense in which Needham will acknowledge that we have learned the composition of water and that water is composed of H_2O. What he would deny is that water *just is* H_2O. Rather, 'the expression "H_2O" describes water's composition and not any of its microstructural features' (2002, p. 223). Beyond dissociation and hydrogen bonding as they are understood, there is still a great deal that chemists do not know about the structure of water.

Hendry resists Needham's approach. We clearly are not determining the extension of 'water' solely by macroscopic qualities when we ask, for example, how much of the human body is water. Rather, we understand such a question to be about a chemical category which need not be clear, tasteless, odorless, and so on. So, Hendry suggests, water is 'the substance that has H_2O molecules as its ingredients'. The molecules will go on to dissociate or arrange themselves by hydrogen bonding, doing a range of things that we only partly understand. Hendry summarizes his proposal as the idea 'that water is the substance formed by bringing together H_2O molecules and allowing them to interact spontaneously' (2006, p. 872). (See also Hendry 2011, ch. 4.)

Of course, we should not understand Hendry to be suggesting that every actual body of water was formed by bringing together H_2O molecules. Particular samples of water may form in a hurly-burly of ionization, such that there is never a moment at which all of the molecules in it are H_2O. It suffices that a substance formed in such a way would be the same substance as the stuff that would form if H_2O molecules were brought together – namely, *water*. To understand Hendry's account in this way, however, we need to rely on a notion of *same stuff*.

We can get the advantages of both Needham's and Hendry's approaches if we think of water as an HPC kind. It is initially identified in terms of macroscopic criteria. Then scientists learn more about its composition and microstructure. The cluster of properties, we learn, is underwritten by a complex causal process. It is not simply a matter of having the right constitution – participating in an essence – but instead of having H_2O, H_3O^+, and OH^- interact in the complex way that they do to form clusters and hydrogen-bonded composites.

The problem with Kornblith's proposal, then, is that it deploys the HPC approach in too simple a way. The 'homeostatic unit' responsible for the properties of water is not merely the bond between two hydrogen atoms and an oxygen atom. That would be taking the metaphor of 'homeostasis' too literally. Rather, what is responsible for the properties of water is the complex causal process that goes on in any sample of water. It is, as Hendry says, the kind of thing that happens when H_2O molecules are brought together.

Note, also, that water does not have just one cluster of macroscopic properties. Like a species, water is polymorphic. Hot water boils into steam. Cold water freezes into ice; there are at least nine different forms of ice with different microstructures. The notion that *water* is preserved through changes in phase was an important innovation of modern chemistry (Needham 2002, § 2). Even though we do not fully understand the causal processes operative in water, we can explain the differences between kinds of ice as differences in crystal structure. This recognition reflects the same three-stage process which I identified in considering a species as an HPC: Begin with a diagnosable cluster of properties, find the causal underpinnings of the cluster, and map the structure of properties resulting from that causal nexus.

E.2 The Unity Problem

In the previous section, I suggested that understanding **water** as an HPC allows us to take the best of Needham's and Hendry's proposals. Yet, water is at most a type-HPC. There are processes going on in the glass of water on my desk, responsible for its possessing a cluster of watery properties. There are processes going on in ice under the Martian surface, responsible for the properties it has. There need not have been any causal connection between the water in this glass and the ice on that distant planet in order for them to be members of the same kind. Although we might worry a bit about the invocation of *sameness* here, we can specify what we mean to a satisfactory degree of precision.

We might go on to try and account for the properties of hydrogen and oxygen atoms in terms of the activities of their subatomic components. Such explanations will crash on the hostile shores of quantum mechanics, where causal stories find no safe harbor. We eventually find components for which there is no further causal story to be given. As Chakravartty notes, the world cannot all be HPCs.

Moreover, the boundaries of type-HPCs will be harder to define in general than the boundaries of token-HPCs. This broaches what Richard Samuels (2009) calls the Unity Problem: determining what it is that makes disparate members of a kind *the same kind of thing*. With a token-HPC, solving the Unity Problem is simply a matter of determining how far the boundaries of the particular causal nexus extend. For a type-HPC, there will be many independent nexuses. Solving the Unity Problem for a type-HPC thus involves determining the boundaries of each particular nexus *and* identifying eligible nexuses wherever they might occur. Members of a type-HPC are in a sense more disparate and so the Unity Problem is more acute.

One might worry that the problem of polymorphic kinds returns. Consider the water on my desk and the ice on Mars. The former is *watery* and the latter *icy*. These are different property clusters. For a sexually dimorphic species, the Unity Problem was solved because a particular individual of one sex relies on members of both sexes for its origin. Paying attention to the cluster of properties typical of females (for example) implicates the cluster typical of males. For developmental polymorphism, the various property clusters are all exhibited in the causal history of an individual organism. In short, the Unity Problem is solved because the initial properties of interest lead us to single token causal process. Such resources might not be available to connect water and ice. Imagine a sample of liquid water which comes together and is electrolyzed without ever freezing. We might study that cluster of wateriness without ever being led to consider ice. The worry comes to this: The underlying causal processes in liquid water and ice are different. Even though **liquid water** and **ice** might be type-HPCs, there is no HPC kind which includes water in both phases. What is similar between liquid water and ice is that they have the same chemical constituents, and so the general kind **water** is better described compositionally than as an HPC.

There are two possible answers to this worry. First, one might argue that the causal processes at work in a sample of liquid water are not so different from the ones at work in a sample of ice. The type of causal process which holds together **liquid water** as an HPC is the same one that holds together **ice**. So they are the same type-HPC after all. I am not sure whether this first reply is successful or not, because the processes responsible for water's wateriness and ice's iciness are similar in some respects but different in others. Sorting this out would require considering the details of water chemistry, which I am not prepared to do. So let's set this reply aside.

Second, we do encounter samples of liquid water which freeze and samples of ice which melt. Paying attention to liquid water, we notice its transformation into ice. That same ice can be melted to produce liquid water again. Although this progression does not always unfold on its own, we both encounter it in the world and can generate it. Just as the development of caterpillars into moths leads us to treat both clusters as part of a property complex, the repeatable phase transitions between liquid water and ice lead us to treat *watery* and *icy* properties as part of a larger complex. There is, of course, a difference. There are no moths that were not first caterpillars, but there may be liquid water which is never ice. This would only be a problem if none of the liquid water *ever* in all history had frozen. Then, perhaps, **liquid water** would be a separate HPC from **ice**. Such a world would be very different than our own, however. I suggest that there would be no singular natural kind **water** in such a world, but just the separate natural kinds **liquid water** and **ice**. The fact that phase transitions do occur, water freezing and melting again, connects the cluster of *watery* properties to the cluster of *icy* ones and so solves the Unity Problem for polymorphic **water**. This resolution only breaks down in exotic possible worlds where phase transitions do not occur, but there is no Unity Problem to solve in such worlds because there is no unified kind.

In this section, I have suggested how **water** might be a type-HPC. Admittedly, my discussion has been somewhat sketchy. One might fill in the details with the cheerful hope that similar work can be done for other chemical kinds as well. In the general case, however, it is a daunting prospect.

F. Coda on HPCs

This chapter has ranged over many topics, and I have argued for a number of distinct but related claims.

Although it does not suffice to define natural kinds as such, an account of HPCs illuminates natural kinds. Many natural kinds are HPCs, and we learn more about them by recognizing this.

We can overcome similarity fetishism by recognizing that kinds of things which are superficially very different are unified by causal relations. The strategy for identifying HPCs begins with the properties of the similar things and charts the boundary of the kind by looking for the causes that are responsible for the properties. The same causes may produce a different but related array of properties in other individuals, and the causal connection leads us to posit a kind which includes all of them. An HPC kind is thus characterized both by the different arrangements of properties which individual members of the kind can possess and the causal processes which maintain the complex arrangement of properties. This would fit poorly with an account of natural kinds which treated them merely as loci of similarity, because the members of an HPC need not all be intrinsically similar. It fits well with the account of natural kinds which I have defended in this book, however. HPCs are real features of the world, and identifying them can be a significant scientific achievement.

Species are an especially clear case. The causal connection between members of the species is participation in the same line of descent or historically related lines; that is, particular species taxa are either what I have called token-HPCs or hybrid-HPCs. Recognizing this, there need be no real dispute between identifying species as historical individuals and identifying them as HPC natural kinds.

Many natural kinds are not connected by a common causal history, and so they are type-HPCs if they are HPCs at all. My somewhat halting remarks about type-HPCs in the final section are not just modest posturing. In writing this chapter of the book, I started with the expectation that chemical kinds would resist any analysis as HPCs. When I learned more about water chemistry, beyond the blithe and familiar formula 'water = H_2O', I began to recognize the complex causal underpinnings required to sustain the superficial features which we tend to associate with water. The three-step process for identifying an HPC which I had described for species was plausibly at work, so I began to see in rough outline how **water** could be seen as an HPC – a type-HPC, but nonetheless a unified kind. I am honestly unsure how far this kind of thinking can be extended: not as far as **electron**, but perhaps further than **water**. Cases cannot be established in the abstract, but instead will depend on the actual structure of things.

7
Conclusion

The theoretical core of this book has been to identify *natural kinds* as those categories which are indispensable for successful science in some specified domain. I have formulated this as two distinct requirements: the *success* clause requires that the category support successful science, and the *restriction* clause requires that it be indispensable. The latter allows us to distinguish the real features of the world from the categories which are introduced merely for scientific bookkeeping; that is, it separates the natural kinds from the merely conventional or fungible kinds. The overall account recognizes natural kinds as categories that support induction and scientific enquiry, categories which arise in and are genuine features of specific domains of objects and phenomena. Scientists themselves are typically only concerned with the first aspect of natural kinds: success. After all, their goal is to do science – not to do metaphysics. (I articulated the core account in Chapter 2, on the basis of desiderata discussed in Chapter 1.)

Many philosophers insist that metaphysics must be about the fundamental categories of being, and so they might say that my conception of natural kinds is not *metaphysical* at all. Such an accusation does not disturb me. Natural kinds as I identify them are real features of the world, even though I do not know whether they are universals, modal structures, or an ontological mélange. Compare the status of particulars like the planet Earth, a duck swimming on a pond, or my cognitive engagement with a jigsaw puzzle. I am comfortable saying that those are real features of the world even though I do not know what they are in their deepest metaphysical depths.

Ignorance about fundamental ontology does not inspire skepticism or nihilism about those things, and there is no reason why it should inspire despair about kinds like **planet**, **duck**, or **cognition** either. (In Chapter 4, I characterized this as a commitment to *equity realism* rather than to *deep realism*.)

However, I am not saying that our enquiry into the ontology of a category should stop once we have identified it as a natural kind. We can ask what it is in the world which structures and sustains the kind, how the kind is held together. Many but not all natural kinds turn out to be Homeostatic Property Clusters (HPCs). Identifying HPCs as what metaphysically underpins (many) natural kinds is not yet to identify a fundamental category, because HPCs might themselves be sustained in various ways. At the very least, we should distinguish token-HPCs (every member of which is historically connected to every other) and type-HPCs (members of which might arise from independent but suitably similar processes). Furthermore, plumbing the ontology of HPCs would require saying more about what *causation* is. Rather than starting with fundamental categories and then putting scientific findings into them as an afterthought, my approach begins from what we know about the world and dives a bit from there. If one refuses to call this 'metaphysics' or to call what I identify 'natural kinds', then one is welcome to substitute different expressions for this *enquiry into what exists* and these *genuine, real categories of stuff and things*. (Chapter 6 discusses many aspects of HPCs and begins the work of identifying the metaphysical underpinnings of some natural kinds.)

Some philosophers might worry about natural kinds being relative to domains of enquiry. Specific domains are only objects of enquiry, after all, because of the contingent fact that scientists decide to enquire into them. This poses a false dichotomy, demanding that kinds which are practical cannot also be real. A sensible outlook acknowledges that we think about what we care about, but also that our caring does not make those things any less real. Often we care – often scientists enquire with such fervor – exactly because the objects of enquiry are real parts of the world. (In Chapter 4, I call this outlook *pragmatic naturalism*.)

An account of natural kinds is not good for much if it only informs abstruse and abstract discussion. A large part of its value is in addressing substantive cases. For a category that we acknowledge or that someone proposes, to what extent does it correspond to a natural kind? In this volume, I have tried to engage a wide range of examples. As the title says: Planets to Mallards. To give a more careful manifest, I remind readers of the kind **planet** and the fate of Pluto (Chapter 3 § A), **water** as a complicated dance of molecules (Chapter 6 § E), species like *Anas platyrhynchos* and the **species** category itself (Chapter 3 § B and Chapter 6 § § A–D), meerkat signs and the threats that they signal (Chapter 5 § C.1), cognition and distributed cognition (Chapter 3 § C), and even baked goods (Chapter 5 § B.2).

Since the details do matter – and since the world is a complicated place – there is more to say about these examples and many more examples to be explored. They can teach us both about the structure of particular domains and about how natural kinds are held together in the world. The back-and-forth between examples and the general account can put theoretical questions into sharper focus. There are topics which I have raised but not resolved: what constitutes a domain, the extent to which natural kinds are a unified or disparate lot, how to establish the bona fide unity of type-HPCs, and so on. As always in philosophy or in science, there is more work to be done.

Bibliography

Ayers, Michael R. (1981), 'Locke versus Aristotle on natural kinds', *Journal of Philosophy* 78(5), 247–72.
Baldwin, James Mark, ed. (1901), *Dictionary of Philosophy and Psychology*, Vol. 1, Macmillan, New York.
Beebee, Helen and Nigel Sabbarton-Leary (2010*a*), On the abuse of the necessary *a posteriori*, in Beebee and Sabbarton-Leary (2010*b*), pp. 159–78.
Beebee, Helen and Nigel Sabbarton-Leary, eds. (2010*b*), *The Semantics and Metaphysics of Natural Kinds*, Routledge, New York.
Bird, Alexander (2010), Discovering the essences of natural kinds, in Beebee and Sabbarton-Leary (2010*b*), pp. 125–36.
Bird, Alexander and Emma Tobin (2009) , Natural kinds, in E. N. Zalta, ed., *The Stanford Encyclopedia of Philosophy*. http://plato.stanford.edu/archives/spr2009/entries/natural-kinds/
Boklage, Charles E. (2006), 'Embryogenesis of chimeras, twins and anterior midline asymmetries', *Human Reproduction* 21(3), 579–91.
Boyd, Richard N. (1982), 'Scientific realism and naturalistic epistemology', *PSA: Proceedings of the Biennial Meeting of the Philosophy of Science Association (1980)* 2, 613–62.
Boyd, Richard N. (1988), How to be a moral realist, in G. Sayre-McCord, ed., *Essays on Moral Realism*, Cornell University Press, Ithaca, New York, pp. 181–228.
Boyd, Richard N. (1989), 'What realism implies and what it does not', *Dialectica* 43(1–2), 5–29.
Boyd, Richard N. (1991), 'Realism, anti-foundationalism and the enthusiasm for natural kinds', *Philosophical Studies* 61, 127–48.
Boyd, Richard N. (1999*a*), Homeostasis, species, and higher taxa, in Wilson (1999), pp. 141–85.
Boyd, Richard N. (1999*b*), Kinds as the 'workmanship of men': Realism, constructivism, and natural kinds, in J. Nida-Rümelin, ed., *Rationalität, Realismus, Revision: Proceedings of the Third International Congress, Gesellschaft für Analytische Philosophie*, Walter de Gruyter, Berlin, pp. 52–89.
Boyd, Richard N. (1999*c*), 'Kinds, complexity and multiple realization', *Philosophical Studies* 95(1/2), 67–98.
Boyd, Richard N. (2010), Realism, natural kinds, and philosophical methods, in Beebee and Sabbarton-Leary (2010*b*), pp. 212–34.
Brigandt, Ingo (2009), 'Natural kinds in evolution and systematics: Metaphysical and epistemological considerations', *Acta Biotheoretica* 59(1–2), 77–97.
Brigandt, Ingo (2010), 'Scientific reasoning is material inference: Combining confirmation, discovery and explanation', *International Studies in the Philosophy of Science* 24(1), 31–43.

Britt, Robert Roy (2006), 'Pluto demoted: No longer a planet in highly controversial definition', *SPACE.com*. Accessed October 29, 2010. http://www.space.com/scienceastronomy/060824_planet_definition.html

Broad, C. D. (1920), 'The relation between induction and probability (part II)', *Mind* 29(113), 11–45.

Brogaard, Berit (2004), 'Species as individuals', *Biology and Philosophy* 19(2), 223–42.

Brookfield, John (2002), 'Review of *Genes, Categories and Species: The Evolutionary and Cognitive Causes of the Species Problem*', *Genetics Research* 79(1), 107–8.

Brown, Alton (2002), *I'm Just Here for the Food*, Stewart, Tabori & Chang, New York.

Brown, Alton (2004), *I'm Just Here for More Food*, Stewart, Tabori & Chang, New York.

Brown, Matt (2011), Science as socially distributed cognition: Bridging philosophy and sociology of science, in K. François, B. Löwe, T. Müller, and B. van Kerkhove, eds., *Foundations of the Formal Sciences VII*, Studies in Logic, College Publications.

Brown, Mike (2006), 'The eight planets'. Posted August 25, 2006, the day after the IAU vote defining 'planet'. http://web.gps.caltech.edu/~mbrown/eightplanets/

Brown, Mike (2010), *How I Killed Pluto and Why It Had It Coming*, Spiegel & Grau, New York.

Cain, Fraser (2008), 'Why Pluto is no longer a planet', *Universe Today*. Accessed October 30, 2010. http://www.universetoday.com/13573/why-pluto-is-no-longer-a-planet/

Cartwright, Nancy (1983), *How the Laws of Physics Lie*, Oxford University Press.

Cartwright, Nancy (1999), *The Dappled World: A Study of the Boundaries of Science*, Cambridge University Press.

Chakravartty, Anjan (2007), *A Metaphysics for Scientific Realism*, Cambridge University Press.

Chaplin, Martin (2010), 'Water structure and science'. Accessed December 14, 2010. http://www.lsbu.ac.uk/water

Chretien, Charles P. (1848), *An Essay on Logical Method*, John Henry Parker, Oxford.

Churchland, Paul M. (1985a), 'Conceptual progress and word/world relations: In search of the essence of natural kinds', *Canadian Journal of Philosophy* 15(1), 1–17.

Churchland, Paul M. (1985b), The ontological status of observables: In praise of the superempirical virtues, in P. M. Churchland and C. A. Hooker, eds., *Images of Science*, University of Chicago Press, pp. 35–47.

Coyne, Jerry A. (2009), *Why Evolution is True*, Viking, New York.

Craver, Carl F. (2009), 'Mechanisms and natural kinds', *Philosophical Psychology* 22(5), 575–94.

Davidson, Donald (1987), 'Knowing one's own mind', *Proceedings and Addresses of the American Philosophical Association* 60(3), 441–58.

de Queiroz, Kevin (1999), The general lineage conception of species and the defining properties of the species category, in Wilson (1999), pp. 49–89.
Dennett, Daniel C. (1991), 'Real patterns', *Journal of Philosophy* 88(1), 27–51.
Dessler, A. J. and C. T. Russell (1980), 'From the ridiculous to the sublime: The pending disappearance of Pluto', *Eos* 61(44), 690.
Devitt, Michael (2008), 'Resurrecting biological essentialism', *Philosophy of Science* 75(3), 344–82.
Douglas, Heather E. (2009), 'Reintroducing prediction to explanation', *Philosophy of Science* 76(4), 444–63.
Douglas, Heather E. (2010), personal communication.
Dupré, John (1989), 'Wilkerson on natural kinds', *Philosophy* 64(248), 248–51.
Dupré, John (1993), *The Disorder of Things: Metaphysical Foundations of the Disunity of Science*, Harvard University Press, Cambridge, Massachusetts.
Dupré, John (1999), On the impossibility of a monistic account of species, in Wilson (1999), pp. 3–22.
Dupré, John (2002), *Humans and Other Animals*, Clarendon Press, Oxford.
Earman, John, Clark Glymour, and John Stachel, eds. (1977), *Minnesota Studies in Philosophy of Science*, Vol. VIII, University of Minnesota Press, Minneapolis.
Eldredge, Niles and Stephen Jay Gould (1972), Punctuated equilibria: An alternative to phyletic gradualism, in T. Schopf, ed., *Models in Paleobiology*, Freeman, Cooper and Company, San Francisco, pp. 82–115.
Elgin, Catherine Z. (2010), 'Keeping things in perspective', *Philosophical Studies* 150(3), 439–47.
Ellis, Brian (2001), *Scientific Essentialism*, Cambridge University Press.
Ereshefsky, Marc (1992), 'Eliminative pluralism', *Philosophy of Science* 59(4), 671–90.
Ereshefsky, Marc (1999), Species and the Linnaean hierarchy, in Wilson (1999), pp. 285–305.
Ereshefsky, Marc (2007), 'Foundational issues concerning taxa and taxon names', *Systematic Biology* 56(2), 295–301.
Ereshefsky, Marc and Mohan Matthen (2005), 'Taxonomy, polymorphism, and history: An introduction to population structure theory', *Philosophy of Science* 72(1), 1–21.
Farber, Ilya (2000), Domain integration: A theory of progress in the scientific understanding of the life and mind, PhD thesis, University of California, San Diego.
Fodor, Jerry Alan (1974), 'Special sciences (or: The disunity of science as a working hypothesis)', *Synthese* 28(2), 97–115.
Franklin, Fabian and Christine Ladd Franklin (1888), 'Mill's natural kinds', *Mind* 13(49), 83–5.
Gelman, Susan A. (1988), 'The development of induction within natural kind and artifact categories', *Cognitive Psychology* 20, 65–95.
Ghiselin, Michael T. (1966), 'On psychologism in the logic of taxonomic controversies', *Systematic Zoology* 15(3), 207–15. http://www.jstor.org/stable/2411392

Ghiselin, Michael T. (1974), 'A radical solution to the species problem', *Systematic Zoology* 23(4), 536–44. http://www.jstor.org/stable/2412471
Ghiselin, Michael T. (1981), 'Categories, life, and thinking', *Behavioral and Brain Sciences* 4(2), 269–83.
Ghiselin, Michael T. (2002), 'Species concepts: The basis for controversy and reconciliation', *Fish and Fisheries* 3(3), 151–60.
Ghiselin, Michael (2007), 'Is the pope a catholic?', *Biology and Philosophy* 22(2), 283–91.
Ghiselin, Michael T. (2009), 'Metaphysics and classification: Update and overview', *Biological Theory* 4(3), 253–9.
Giere, Ronald N. (2002), Scientific cognition as distributed cognition, in P. Carruthers, S. Stitch, and M. Siegal, eds., *The Cognitive Basis of Science*, Cambridge University Press, pp. 285–99.
Giere, Ronald N. (2006), *Scientific Perspectivism*, University of Chicago Press, Chicago.
Goode, George Brown and Tarleton H. Beane (1895), *Oceanic Ichthyology based on the study of Deep-sea Fishes of the Atlantic Basin*, Government Printing Office, Washington, D.C.
Goodman, Nelson (1947), 'The problem of counterfactual conditionals', *Journal of Philosophy* 44(5), 113–28.
Goodman, Nelson (1978), *Ways of Worldmaking*, Hackett Publishing Company, Indianapolis.
Goodman, Nelson (1983 [1954]), *Fact, Fiction, and Forecast*, 4th edn., Harvard University Press, Cambridge, Massachusetts.
Griffiths, Paul E. (1999), Squaring the circle: Natural kinds with historical essences, in Wilson (1999), pp. 209–28.
Hacking, Ian (1983), *Representing and Intervening*, Cambridge University Press.
Hacking, Ian (1986), Making up people, in T. C. Heller, M. Sosna, and D. E. Wellbery, eds., *Reconstructing Individualism*, Stanford University Press, pp. 222–36.
Hacking, Ian (1991a), 'On Boyd', *Philosophical Studies* 61, 149–54.
Hacking, Ian (1991b), 'A tradition of natural kinds', *Philosophical Studies* 61, 109–26.
Hacking, Ian (1995), The looping effects of human kinds, in D. Sperber, D. Premack, and A. J. Premack, eds., *Causal Cognition*, Clarendon Press, Oxford, pp. 351–83.
Hacking, Ian (1999), *The Social Construction of What?*, Harvard University Press, Cambridge, Massachusetts.
Hacking, Ian (2007a), 'The contingencies of ambiguity', *Analysis* 67(4), 269–77.
Hacking, Ian (2007b), 'Natural kinds: Rosy dawn, scholastic twilight', *Royal Institute of Philosophy Supplement* 82, 203–39.
Hacking, Ian (2007c), 'Putnam's theory of natural kinds and their names is not the same as Kripke's', *Principia* 11(1), 1–24.
Hempel, Carl G. (1961), Problems of taxonomy and their application to nosology and nomenclature in the mental disorders, in J. Zubin, ed., *Field*

Studies in Mental Disorders, Grune & Stratton, New York, pp. 3–50, with transcript of discussion.

Hempel, Carl G. (1965a), Fundamentals of taxonomy, in *Aspects of Scientific Explanation and other Essays in the Philosophy of Science*, The Free Press, New York, pp. 137–54.

Hempel, Carl G. (1965b), *Aspects of Scientific Explanation and other Essays in the Philosophy of Science*, The Free Press, New York.

Hendry, Robin Findlay (2006), 'Elements, compounds and other chemical kinds', *Philosophy of Science* 73(5), 864–75.

Hendry, Robin Findlay (2010), The elements and conceptual change, in Beebee and Sabbarton-Leary (2010b), pp. 137–58.

Hendry, Robin Findlay (2012), *The Metaphysics of Chemistry*, Oxford University Press. Forthcoming.

Hull, David L. (1976), 'Are species really individuals?', *Systematic Zoology* 25(2), 174–91.

Hull, David L. (1978), 'A matter of individuality', *Philosophy of Science* 45(3), 335–60.

Hutchins, Edwin (1995), *Cognition in the Wild*, MIT Press, Cambridge, Massachusetts.

Hutchins, Edwin (2008), 'The role of cultural practices in the emergence of modern human intelligence', *Proceedings of the Royal Society B* 363(1499), 2011–19.

IAU (2006), 'Resolution B5: Definition of a planet in the solar system'. Accessed October 30, 2010. http://www.iau.org/static/resolutions/Resolution_GA26-5-6.pdf

IAU (2010), 'List of Jupiter trojans', IAU minor planet center. Accessed October 30, 2010. http://www.cfa.harvard.edu/iau/lists/JupiterTrojans.html

Isaac, Nick J. B., James Mallet, and Georgina M. Mace (2004), 'Taxonomic inflation: Its influence on macroecology and conservation', *Trends in Ecology and Evolution* 19(9), 464–9.

IUPAC (1997), *Compendium of Chemical Terminology, the 'Gold Book'*, 2nd edn., Blackwell Scientific Publications, Oxford. Compiled by A. D. McNaught and A. Wilkinson. XML on-line corrected version: http://goldbook.iupac.org (2006–) created by M. Nic, J. Jirat, B. Kosata; updates compiled by A. Jenkins.

James, William (1948), The will to believe, in A. Castell, ed., *Essays in Pragmatism*, Hafner Publishing Co., New York, pp. 88–109. Originally published June, 1896.

Kahane, Howard (1969), 'Thomason on natural kinds', *Noûs* 3(4), 409–12.

Keller, Roberto A., Richard N. Boyd, and Quentin D. Wheeler (2003), 'The illogical basis of phylogenetic nomenclature', *The Botanical Review* 69(1), 93–110.

Kellert, Stephen H., Helen E. Longino, and C. Kenneth Waters (2006), Introduction: The pluralist stance, in S. H. Kellert, H. E. Longino, and C. K. Waters, eds., *Minnesota Studies in Philosophy of Science*, Vol. XIX, University of Minnesota Press, Minneapolis, pp. vii–xxix.

Khalidi, Muhammad Ali (1993), 'Carving nature at the joints', *Philosophy of Science* 60(1), 100–13.
Kitcher, Philip (1984), 'Species', *Philosophy of Science* 51(2), 308–33.
Kitcher, Philip (1984/1985), 'Good science, bad science, dreadful science, and pseudoscience', *Journal of College Science Teaching* 14(3), 168–73.
Kitcher, Philip (1993), *The Advancement of Science*, Oxford University Press.
Kitcher, Philip (2001), *Science, Truth, and Democracy*, Oxford University Press.
Kitcher, Philip (2002), 'On the explanatory role of correspondence truth', *Philosophy and Phenomenological Research* 64(2), 346–64.
Kornblith, Hilary (1993), *Inductive Inference and Its Natural Ground*, MIT Press, Cambridge, Massachusetts.
Kripke, Saul A. (1972), *Naming and Necessity*, Harvard University Press, Cambridge, Massachusetts.
Laamanen, T. R., F. T. Petersen, and R. Meier (2003), 'Kelp flies and species concepts – the case of *Coelopa frigida* (Fabricius, 1805) and *C. nebularum* Aldrich, 1929 (Diptera: Coelopidae)', *Journal of Zoological Systematics and Evolutionary Research* 41(2), 127–36.
Ladyman, James (1998), 'What is structural realism?', *Studies in the History and Philosophy of Science* 29(3), 409–24.
Ladyman, James and Don Ross with David Spurrett and John Collier (2007), *Everything Must Go: Metaphysics Naturalized*, Oxford University Press.
LaPorte, Joseph (2004), *Natural Kinds and Conceptual Change*, Cambridge University Press.
LaPorte, Joseph (2010), Theoretical identity statements, their truth, and their discovery, in Beebee and Sabbarton-Leary (2010*b*), pp. 104–24.
Laudan, Larry (1990), Demystifying underdetermination, in C. W. Savage, ed., *Minnesota Studies in Philosophy of Science*, Vol. XIV, University of Minnesota Press, Minneapolis, pp. 267–97.
Lawniczak, M. K. N., S. J. Emrich, A. K. Holloway, A. P. Regier, M. Olson, B. White, S. Redmond, L. Fulton, E. Appelbaum, J. Godfrey, C. Farmer, A. Chinwalla, S.-P. Yang, P. Minx, J. Nelson, K. Kyung, B. P. Walenz, E. Garcia-Hernandez, M. Aguiar, L. D. Viswanathan, Y.-H. Rogers, R. L. Strausberg, C. A. Saski, D. Lawson, F. H. Collins, F. C. Kafatos, G. K. Christophides, S. W. Clifton, E. F. Kirkness, and N. J. Besansky (2010), 'Widespread divergence between incipient *Anopheles gambiae* species revealed by whole genome sequences', *Science* 330(6003), 512–14.
Lehmann, Tovi and Abdoulaye Diabate (2008), 'The molecular forms of *Anopheles gambiae*: A phenotypic perspective', *Infection, Genetics and Evolution* 8(5), 737–46.
Leonard, Henry S. and Nelson Goodman (1940), 'The calculus of individuals and its uses', *Journal of Symbolic Logic* 5(2), 45–55.
Lowe, E. J. (2009), 'The rationality of metaphysics', *Synthese* 178(1), 99–109.
Machery, Edouard (2012), 'Why I stopped worrying about the definition of life ... and why you should as well', *Synthese* 185(1), 145–64.
Mackie, J. L. (1974), 'Locke's anticipation of Kripke', *Analysis* 34(6), 177–80.

Mag Uidhir, Christy and P. D. Magnus (2011), 'Art concept pluralism', *Metaphilosophy* 42(1-2), 83-97.

Magnus, P. D. (2003), Underdetermination and the claims of science, PhD thesis, University of California, San Diego. http://hdl.handle.net/1951/42590

Magnus, P. D. (2005a), 'Background theories and total science', *Philosophy of Science* 75(2), 1064-75.

Magnus, P. D. (2005b), 'Peirce: Underdetermination, agnosticism, and related mistakes', *Inquiry* 48(1), 26-37.

Magnus, P. D. (2007), 'Distributed cognition and the task of science', *Social Studies of Science* 37(2), 297-310.

Magnus, P. D. (2008), 'Demonstrative induction and the skeleton of inference', *International Studies in the Philosophy of Science* 22(3), 303-15.

Magnus, P. D. (2010), 'Inductions, red herrings, and the best explanation for the mixed record of science', *British Journal for the Philosophy of Science* 61(4), 803-19.

Magnus, P. D. (2011a), 'Drakes, seadevils, and similarity fetishism', *Biology & Philosophy* 26(6), 857-70.

Magnus, P. D. (2011b), 'Miracles, trust, and ennui in Barnes' *Predictivism*', *Logos & Episteme* 2(1), 103-115. http://logos-and-episteme.proiectsbc.ro/?q=node/73

Magnus, P. D. (2012), Historical individuals like *Anas platyrhynchos* and 'Classical gas', in C. Mag Uidhir, ed., *Art and Abstract Objects*, Oxford University Press.

Magnus, P. D. and Craig Callender (2004), 'Realist ennui and the base rate fallacy', *Philosophy of Science* 71(3), 320-38.

Malament, David (1977), Observationally indistinguishable space-times, in J. Earman, C. Glymour, and J. Stachel, eds., *Minnesota Studies in Philosophy of Science*, Vol. VIII, University of Minnesota Press, Minneapolis, pp. 61-80.

Manchak, John Byron (2009), 'Can we know the global structure of space-time?', *Studies in History and Philosophy of Modern Physics* 40(1), 53-6.

Manser, Marta B. (2001), 'The acoustic structure of suricates' alarm calls varies with predator type and the level of response urgency', *Proceedings of the Royal Society B* 268(1483), 2315-24.

Manser, Marta B., Matthew B. Bell, and Lindsay B. Fletcher (2001), 'The information that receivers extract from alarm calls in suricates', *Proceedings of the Royal Society B* 268(1484), 2485-91.

Manser, Marta B., Robert M. Seyfarth, and Dorothy L. Cheney (2002), 'Suricate alarm calls signal predator class and urgency', *TRENDS in Cognitive Sciences* 6(2), 55-7.

Marr, David (1982), *Vision*, W. H. Freeman and Company, San Francisco.

Martin, Robert M. (2000), *Scientific Thinking*, Broadview Press, Peterborough, Ontario.

Martineau, James (1859), 'John Stuart Mill', *The National Review* 9(18), 474-508. A review of Mill's *Dissertations and Discussions, Political, Philosophical, and Historical*.

Mayden, Richard L. (1997), A hierarchy of species concepts: The denouement in the saga of the species problem, in *Species: The Units of Biodiversity*, Chapman Hall, London, pp. 381–424.
Mayden, Richard L. (1999), 'Consilience and a hierarchy of species concepts: Advances toward closure on the species puzzle', *Journal of Nematology* 31(2), 95–116.
Mayr, Ernst (1964 [1942]), *Systematics and the Origin of Species from the Viewpoint of a Zoologist*, Dover, New York.
McCaig, Donald (2007), *The Dog Wars*, Outrun Press, Hillsborough, New Jersey.
McClamrock, Ron (1991), 'Marr's three levels: A re-evaluation', *Minds and Machines* 1(2), 185–96.
McClamrock, Ron (1995), *Existential Cognition: Computational Minds in the World*, University of Chicago Press.
McClelland, James L., David E. Rumelhart, and the PDP Research Group (1986), *Parallel Distributed Processing*, Vol. 2, MIT Press, Cambridge, Massachusetts.
McOuat, Gordon (2009), 'The origins of "natural kinds": Keeping "essentialism" at bay in the age of reform', *Intellectual History Review* 19(2), 211–30.
Mill, John Stuart (1874), *A System of Logic*, 8th edn., Harper Brothers, New York.
Millikan, Ruth Garrett (1984), *Language, Thought, and Other Biological Categories: New Foundations for Realism*, MIT Press, Cambridge, Massachusetts.
Millikan, Ruth Garrett (1996), 'On swampkinds', *Mind & Language* 11(1), 103–17.
Millikan, Ruth Garrett (1999), 'Historical kinds and the "special sciences"', *Philosophical Studies* 95(1/2), 45–65.
Mishler, Brent D. (1999), Getting rid of species?, in Wilson (1999), pp. 307–15.
Miya, Masaki, Theodore W. Pietsch, James W. Orr, Rachel J. Arnold, Takashi P. Satoh, Andrew M. Shedlock, Hsuan-Ching Ho, Mitsuomi Shimazaki, Mamoru Yabe, and Mutsumi Nishida (2010), 'Evolutionary history of anglerfishes (teleostei: Lophiiformes): A mitogenomic perspective', *BMC Evolutionary Biology* 10(58), 1–27.
Mumford, Stephen (2005), 'Kinds, essences, powers', *Ratio* 18(4), 420–36.
NASA (2004), 'Astronomy picture of the day: 2004 May 14 - Zubenelgenubi and friends'. Accessed August 24, 2011. http://apod.nasa.gov/apod/ap040514.html
Neafsey, D. E., M. K. N. Lawniczak, D. J. Park, S. N. Redmond, M. B. Coulibaly, S. F. Traorž, N. Sagnon, C. Costantini, C. Johnson, R. C. Wiegand, F. H. Collins, E. S. Lander, D. F. Wirth, F. C. Kafatos, N. J. Besansky, G. K. Christophides, and M. A. T. Muskavitch (2010), 'SNP genotyping defines complex gene-flow boundaries among African malaria vector mosquitoes', *Science* 330(6003), 514–17.
Needham, Paul (2000), 'What is water?', *Analysis* 60(1), 13–21.
Needham, Paul (2002), 'The discovery that water is H_2O', *International Studies in the Philosophy of Science* 16(3), 205–26.

Newburgh, Ronald, Joseph Peidle, and Wolfgang Rueckner (2006), 'Einstein, Perrin, and the reality of atoms: 1905 revisited', *American Journal of Physics* 74(6), 478–81.

Norton, John D. (2003), 'A material theory of induction', *Philosophy of Science* 70(4), 647–70.

Page, Sam (2006), 'Mind-independence disambiguated: Separating the meat from the straw in the realism/anti-realism debate', *Ratio* 19(3), 321–35.

Pauling, Linus (1970), *General Chemistry*, 3rd edn., W. H. Freeman, San Francisco.

Peirce, Charles Sanders (1992 [1878]), How to make our ideas clear, in N. Houser and C. Kloesel, eds., *The Essential Peirce*, Vol. 1, Indiana University Press, Bloomington, pp. 124–141.

Pietsch, Theodore W. (2005), 'Dimorphism, parasitism, and sex revisited: Modes of reproduction among deep-sea ceratoid anglerfishes', *Ichthyological Research* 52, 207–36.

Pietsch, Theodore W. (2009), *Oceanic Anglerfishes: Extraordinary Biodiversity in the Deep Sea*, University of California Press, Berkeley, California.

Pietsch, Theodore W. and Christopher P. Kenaley (2007), 'Ceratioidei sea-devils, devilfishes, deep-sea anglerfishes'. Part of *The Tree of Life Web Project*, http://tolweb.org/. Accessed September 15, 2010. http://tolweb.org/Ceratioidei/22000/2007.10.02

Pincock, Christopher (2010), 'Exploring the boundaries of conceptual evaluation', *Philosophia Mathematica* 18(1), 106–21. Review of Mark Wilson's *Wandering Significance*.

Popper, Karl Raimund (1963, 1965), *Conjectures and Refutations*, Harper & Row, New York.

Popper, Karl Raimund (1966), *The Open Society and Its Enemies*, Vol. I, 5th edn., Princeton University Press, Princeton, New Jersey.

Psillos, Stathis (2002), *Causation & Explanation*, McGill-Queen's University Press, Montreal.

Putnam, Hilary (1975a [1970]), Is semantics possible?, in *Mind, Language, and Reality: Philosophical Papers Volume 2*, Cambridge University Press, pp. 139–52.

Putnam, Hilary (1975b), The meaning of 'meaning', in K. Gunderson, ed., *Minnesota Studies in Philosophy of Science*, Vol. VII, University of Minnesota Press, Minneapolis, pp. 131–93.

Putnam, Hilary (1981), *Reason, Truth and History*, Cambridge University Press.

Quine, Willard Van Orman (1969a), Natural kinds, in Quine (1969b), pp. 114–38.

Quine, Willard Van Orman (1969b), *Ontological Relativity & other essays*, Columbia University Press, New York.

Read, Carveth (1878), *On the Theory of Logic: An Essay*, C. Kegan Paul & Co., London.

Reid, Thomas (1997 [1764]), *An Inquiry into the Human Mind on the Principles of Common Sense*, Pennsylvania State University Press, University Park, Pennsylvania. Critical edition, edited by Derek R. Brookes.

Reisch, George A. (1998), 'Pluralism, logical empiricism, and the problem of pseudoscience', *Philosophy of Science* 65(2), 333–48.

Reydon, Thomas A. C. (2003), 'Species are individuals: Or are they?', *Philosophy of Science* 70(1), 49–56.

Reydon, Thomas A. C. (2008), 'Species in three and four dimensions', *Synthese* 164(2), 161–84.

Reydon, Thomas A. C. (2010), 'Classifying artifacts is like classifying people: Looping effects for artifact kinds'. Unpublished manuscript.

Richards, Richard A. (2010), *The Species Problem: A Philosophical Analysis*, Cambridge University Press.

Rorty, Richard (1999), *Philosophy and Social Hope*, Penguin Books, London.

Rouse, G. W., S. K. Goffredi, and R. C. Vrijenhoek (2004), 'Osedax: Bone-eating marine worms with dwarf males', *Science* 305(5684), 668–71. http://www.sciencemag.org/cgi/content/abstract/305/5684/668

Rouse, G. W., K. Worsaae, S. B. Johnson, W. J. Jones, and R. C. Vrijenhoek (2008), 'Acquisition of dwarf male "harems" by recently settled females of Osedax roseus n. sp. (Siboglinidae; Annelida)', *Biological Bulletin* 214(1), 67–82. http://www.jstor.org/stable/25066661

Rouse, Greg, Nerida Wilson, Shana Goffredi, Shannon Johnson, Tracey Smart, Chad Widmer, Craig Young, and Robert Vrijenhoek (2009), 'Spawning and development in Osedax boneworms (siboglinidae, annelida)', *Marine Biology* 156, 395–405.

Russell, Bertrand (1948), *Human Knowledge: Its Scope and Limits*, Simon & Schuster, New York.

Samuels, Richard (2009), Delusions as a natural kind, in M. R. Broome and L. Bortolotti, eds., *Psychiatry as Cognitive Neuroscience: Philosophical Perspectives*, Oxford University Press, pp. 49–82.

Samuels, Richard and Michael Ferreira (2010), 'Why don't concepts constitute a natural kind?', *Behavioral and Brain Sciences* 33, 222–3.

Schwartz, Robert (2000), 'Starting from scratch: Making worlds', *Erkenntnis* 52(2), 151–9.

Sklar, Lawrence (2003), 'Dappled theories in a uniform world', *Philosophy of Science* 70(2), 424–41.

Slater, Matthew H. (unpublished), 'Pluto and the platypus: Tale of an odd ball and an odd duck'.

Stanford, P. Kyle (2006), *Exceeding Our Grasp: Science, History, and the Problem of Unconceived Alternatives*, Oxford University Press.

Stroll, Avrum (1998), *Sketches of Landscapes*, MIT Press, Cambridge, Massachusetts.

Tan, Denise S. H., Farhan Ali, Sujatha Narayanan Kutty, and Rudolf Meier (2008), 'The need for specifying species concepts: How many species of silvered langurs (*Trachypithecus cristatus* group) should be recognized?', *Molecular Phylogenetics and Evolution* 47(2), 629–36.

Tan, Denise S. H., Yuchen Ang, Gwynne S. Lim, Mirza Rifqi Bin Ismail, and Rudolf Meier (2010), 'From "cryptic species" to integrative taxonomy: An iterative process involving DNA sequences, morphology, and behaviour leads to the resurrection of *Sepsis pyrrhosoma* (Sepsidae: Diptera)', *Zoologica Scripta* 39(1), 51–61.

Tattersall, Ian (2007), 'Madagascar's lemurs: Cryptic diversity or taxanomic inflation', *Evolutionary Anthropology* 16(1), 12–23.

Thomasson, Amie L. (2003), 'Realism and human kinds', *Philosophy and Phenomenological Research* 67(3), 580–609.

Tobin, Emma (2010), Crosscutting natural kinds and the hierarchy thesis, in Beebee and Sabbarton-Leary (2010*b*), pp. 179–91.

Towry, M. H. (1887), 'On the doctrine of natural kinds', *Mind* 12(47), 434–8.

Unger, Peter (1980), 'The problem of the many', *Midwest Studies in Philosophy* V, 411–67.

van Fraassen, Bas C. (1980), *The Scientific Image*, Clarendon Press, Oxford.

Venn, John (1866), *The Logic of Chance*, Macmillan and Co., London and Cambridge.

Weatherson, Brian (2009), The problem of the many, in E. N. Zalta, ed., *The Stanford Encyclopedia of Philosophy*. http://plato.stanford.edu/archives/win2009/entries/problem-of-many/

Wegner, Daniel M. (1986), Transactive memory: A contemporary analysis of the group mind, in B. Mullen and G. R. Goethals, eds., *Theories of Group Behavior*, Springer-Verlag, New York, pp. 185–208.

Wegner, Daniel M., Toni Giuliano, and Paula T. Hertel (1985), Cognitive interdependence in close relationships, in W. Ickes, ed., *Compatible and Incompatible Relationships*, Springer-Verlag, New York, pp. 253–76.

Weintraub, David A. (2007), *Is Pluto a Planet?*, Princeton University Press, Princeton, New Jersey.

Wilkerson, Terence Edward (1988), 'Natural kinds', *Philosophy* 63(243), 29–42.

Wilkerson, Terence Edward (1993), 'Species, essences and the names of natural kinds', *Philosophical Quarterly* 43(170), 1–19.

Wilkerson, Terence Edward (1995), *Natural Kinds*, Avebury, Aldershot.

Wilkins, John S. (2009), *Species: A History of the Idea*, University of California Press, Berkeley, California.

Wilson, Mark (1982), 'Predicate meets property', *Philosophical Review* 91(4), 549–89.

Wilson, Mark (1985), 'What is this thing called "pain"? The philosophy of science behind the contemporary debate', *Pacific Philosophical Quarterly* 66(3–4), 227–67.

Wilson, Mark (2006), *Wandering Significance*, Clarendon Press, Oxford.

Wilson, Robert A., ed. (1999), *Species: New Interdisciplinary Essays*, MIT Press, Cambridge, Massachusetts.

Wilson, Robert A. (2004), *Boundaries of the Mind*, Cambridge University Press.

Wilson, Robert A. (2005), *Genes and the Agents of Life*, Cambridge University Press.
Wilson, Robert A., Matthew J. Barker, and Ingo Brigandt (2007), 'When traditional essentialism fails: Biological natural kinds', *Philosophical Topics* 35, 189–215.
Worrall, John (1989), 'Structural realism: The best of both worlds?', *Dialectica* 43(1–2), 99–124.
Worsaae, K. and G. W. Rouse (2010), 'The simplicity of males: Dwarf males of four species of Osedax (Siboglinidae; Annelida) investigated by confocal laser scanning microscopy', *Journal of Morphology* 271, 127–42.
Zhang, Jiajie and Donald A. Norman (1995), 'A representational analysis of numeration systems', *Cognition* 57, 271–95.

Index

amphibolic pragmatism, 109–13
anglerfish, 40, 88, 90, 159, 160–3
artifacts, 23–6, 64
automobiles, 24–5

baked goods, 23, 133–6, 145–6
Barker, Matthew, 154
Bendorf, Adam, ix
Bennett, Karen, ix
Birch, Jonathan, x
Bird, Alexander, 31, 40, 59
Bocklage, Charles, 35
Bogen, Jim, x
border collies, 63–5
Boyd, Richard, ix, 6, 10, 20, 27–8, 39, 41–3, 52–3, 105–9, 137, 149–51, 153–4, 156–7, 159, 162, 168–72, 175, 183
Brigandt, Ingo, 42, 154, 156
Broad, C. D., 151
Brogaard, Berit, 177–8
Brookfield, John, 83–4
Brown, Alton, 133–6
Brown, Matthew, ix, 101–2
Brown, Mike, 72, 75, 81–3
Bursten, Julia, x

Cartwright, Nancy, 21–2, 123
Chakravartty, Anjan, 124–5, 127, 129, 183–5, 188
chemistry, 28, 57–61
  see also: gold, water
chimeras, 35
Churchland, Paul, ix, 60, 104, 114, 139–40
clades, 89–90
colloids, 28
constellations, 108–9, 142–5
Coyne, Jerry, 25

Craver, Carl, 163–4
creationism, 130–2

Danks, David, x
Darmstadtium, 25–6, 141
Davidson, Donald, 166
de Queiroz, Kevin, 92–4
deep realism, 119, 122–5, 132, 136
Deinhart, Brian, x
DeLeo, Chris, x
Devitt, Michael, 169
distributed cognition, 96–102
dogs, 25–6, 63–5, 91, 104, 176–80
domains of enquiry, 43–5
double-blind trials, 100–1
Douglas, Heather, x
Duhem–Quine problem, 51–2
Dupré, John, ix, 34–6, 95–6, 106, 126, 129–33
dwarf planet, 80–1

Eklund, Matti, ix
Eldredge, Niles, 159, 165
electrons, 12, 147, 184
Elgin, Catherine, 104
Ellis, Brian, 6, 26–7
equity realism, 20, 119–22
Ereshefsky, Marc, 89, 93, 95, 130–1, 151–6
essences, 18–19, 28, 32–6, 117–19, 166, 185

Farber, Ilya, 52
Farreira, Michael, 147
Feuer, Daniel, x
Fodor, Jerry, 36
Frost-Arnold, Greg, ix
functional kinds, 36–7, 112–13
fungible kinds, 61–7, 134, 141–2, 142–3, 145

Gelman, Susan, 23–4
Ghiselin, Michael, 94, 175, 177
Giere, Ron, 98–101
gold, 32–3, 41–2
Goodman, Nelson, 9–10, 22–3, 180–1
Gould, Stephen Jay, 159, 165
Griffiths, Paul, 13–16, 150
grue, 9–10

Hacking, Ian, 5–7, 29, 31, 64–5, 123
Hempel, Carl, 25–6
Hendry, Robin, ix, 29, 58, 133, 186–8
hierarchy assumption, 14, 37–8
homeostatic property clusters, 109, 147–91, 193
Hull, David, 175, 182–3
Hutchins, Edwin, 108, 144

induction assumption, 8–18, 24, 33, 38, 55–6, 160–5
instrumentation, 65–7
intrinsic properties, 32–7, 69

jackal poop, 138–9
jade, 12, 31–2
James, William, 116

Kahane, Howard, 38
Kellert, Stephen, 130
Khalidi, Muhammad Ali, 38
Khalifa, Kareem, x
Kitcher, Philip, ix, 122, 131–2, 169–72, 177–9, 182
Kornblith, Hilary, 6, 185, 187
Kripke, Saul, 32
Kripke–Putnam view of reference, 6, 29–32, 69–70, 107, 115, 150

Ladyman, James, 123, 130
Lagrange Points, 75–7
Laporte, Joseph, 30–1
Laudan, Larry, 50
laws of nature, 19, 21–3, 69
lemurs, 92

Leonard, Henry, 180
Leuridan, Bert, x
life, 53–4, 175
Linnaean system, 37, 91, 95
Locke, John, 6–7, 18–19
Longino, Helen, 130
looping effects, 64–5
Lowe, E. J., 132

Machery, Eduoard, x, 53–4
Malament, David, 114
mallards, 15, 149–60, 169–71, 172, 181, 183
Manchak, John, 114
Manser, Marta, 137–9
Marr, David, 96–7
Martin, Robert, 103–4, 121
material theory of induction, 42
Matthen, Mohan, 151–6
Mayden, Richard, 84, 92–4
Mayr, Ernst, 84, 89
McCaffrey, Joseph, x
McClamrock, Ron, 97
McOuat, Gordon, 6, 18
meerkats, 136–40
Milanese, John, x
Mill, John Stuart, 5–6, 10, 16–18
Millikan, Ruth Garrett, 10, 24–5, 168–9
Mishler, Brent, 95
Mitchell, Sandra, ix, x
mosquitoes, 85–7, 91, 95, 129, 172–4
Mumford, Stephen, 125, 130

Needham, Paul, 186–8
the No-Miracles Argument, 121–3
Norman, Donald, 98–9
Norton, John, x, 42

Odenbaugh, Jay, ix
Osedax, 66–7, 162–3
oxygen, 60–1, 106–8

Page, Sam, 49, 111–13, 142
Pauling, Linus, 59

Peirce, Charles Sanders, 5–6, 17, 113–14
peppered moths, 152–4, 189
phases of matter, 37–8, 188–90
phlogiston, 60–1
phosphorus, 17–18
Pietsch, Theodore, 160–1, 163
Pincock, Christopher, 117
planets, 1–2, 68–83
Plato, 1
Popper, Karl, 18, 131
pragmatic naturalism, 103, 118–20, 193
problem of the many, 112
promiscuous realism, 126, 129–36
Psillos, Stathis, 21
Putnam, Hilary, 6, 29–32, 34, 39, 113, 166, 185
  see also: the Kripke–Putnam view of reference

Quine, Willard Van Orman, 5–6, 11–12, 17, 41, 151

Read, Carveth, 40
realism
  entity realism, 123
  internal realism, 113–15
  metaphysical realism, 39
  scientific realism, 103–4, 119–23
  structural realism, 123
  see also: deep realism, equity realism, promiscuous realism, semirealism
Redies, Tiffany, x
Reid, Thomas, 5
Reisch, George, 130–1
restriction clause, 48–50, 56–9, 62, 66–7, 68, 78, 108–9, 121–2, 129, 192
Reydon, Thoman, 65, 181–2
Richards, Richard, 84, 92–4, 178
rigid designation, 29–32, 69–70
Rorty, Richard, 109–11
Ross, Don, 123, 130
Russell, Bertrand, 5–6, 124, 178

Samuels, Richard, x, 147, 170, 188–90
the scarcity assumption, 38–9, 103–4
schnatural schkinds, 4, 193
Schwartz, Robert, 143–5
scientific success, 47–8
semirealism, 124, 127
sharpness, 19
Silins, Nico, ix
similarity fetishism, 11–12, 55, 156–65, 190
the simpliciter assumption, 39–43, 54–5, 114–15
Sklar, Lawrence, 22
Slater, Matthew, 83
species, 34–6, 37, 83–96, 130–3
  individualism about species, 84–5, 175–83
  realism about species, 84, 85–7, 95, 172
  species category pluralism, 87–96, 172–5
Stanford, Kyle, 120
Stroll, Avrum, 30
success clause, 47–50, 192
swampman, 166–7

Tattersall, Ian, 92
Thibeault, Damian, x
Thomasson, Amie, 135
tigers, 32, 166–7
Tobin, Emma, 38, 40
transactive memory, 100
Trojan asteroids, 75–7

underdetermination of theory by data, ix, 50–5
unicorns, 40, 48, 52, 141–2
the unity problem, 188–90

van Fraassen, Bas, 119–20
Vickers, Peter, x

water, 30–2, 104–5, 184–91
Waters, Kenneth, 130
Wegner, Daniel, 100

Weintraub, David, 72, 75–7
Weisberg, Michael, ix
whales, whether they are fish, 7
wholesale vs. retail arguments, 120–3
Wilkerson, T. E., 21, 28, 32–6, 127–8, 132, 167–8
Wilkins, John, 18, 84, 87

Wilson, Mark, 31–2, 36–7, 41, 109, 115–17
Wilson, Robert, 97, 154, 172–5
Woodward, Jim x
Worrall, John, 123

Zhang, Jiajie, 98–9